Molecular Structure Description

THE ELECTROTOPOLOGICAL STATE

Molecular Structure Description

THE ELECTROTOPOLOGICAL STATE

Lemont B. Kier

Virginia Commonwealth University
Richmond, Virginia

Lowell H. Hall

Eastern Nazarene College
Quincy, Massachusetts

ACADEMIC PRESS

01-004

San Diego   London   Boston   New York   Sydney   Tokyo   Toronto

Front cover photograph: The E-State and HE-State 3-D fields were calculated using Molconn-Z 3.10S and contoured and displayed by Sybyl graphics on a Silicon Graphics, Inc. workstation by Glen E. Kellogg (Edusoft LC, Ashland, VA). Polar hydrogen atoms are indicated by large green contours; atoms with high electron accessibility are indicated by large orange contours. Sybyl is a product of Tripos, Inc., St. Louis, MO.

This book is printed on acid-free paper. ∞

Academic Press
a division of Harcourt Brace & Company
525 B Street, Suite 1900, San Diego, California 92101-4495, USA
http://www.apnet.com

Academic Press
24-28 Oval Road, London NW1 7DX, UK
http://www.hbuk.co.uk/ap/

Library of Congress Catalog Card Number: 98-89313

International Standard Book Number: 0-12-406555-4

PRINTED IN THE UNITED STATES OF AMERICA
99 00 01 02 03 04 ML 9 8 7 6 5 4 3 2 1

We dedicate this book to the memories
of our doctoral advisors and major professors:

Taito O. Soine (LBK) and Walter S. Koski (LHH).

The time spent with a capable, generous, and sensitive mentor
can shape a productive and rewarding career.
We are grateful for those early experiences.

CONTENTS

3 Significance and Interpretations of the E-State Indices

4 Extended Forms of the E-State

5 Strategies for Use of the E-State

CD contains the software E-Calc, the User's Guide, and online help.

FOREWORD

Over the years, chemical scientists have sought to develop indices reflective of the local and global properties of molecules. To my mind, however, very few of these indices have the intuitive appeal and generality of the electrotopological state indices developed by Monty Kier and Lowell Hall nearly 10 years ago. Electrotopological states, or E-States, are fictitious atomic states—not quantum states—that characterize an atom's properties due to its local and global molecular environments.

Local atom properties are defined by what Kier and Hall call an I-State, which is related to a valence state. Valence states are not quantum mechanically well-defined atomic states. Rather, they correspond to pseudostates that are based on an atom's hybridization and its electronic configuration within a given molecular environment. For example, the electronic configuration of an sp^3 carbon valence state places each of the carbon's four valence electrons in one of its four tetrahedrally directed sigma orbitals: $(td)^4$. In contrast, the electronic configuration of an sp^2 carbon valence state places one electron in each of its three trigonal sigma orbitals and one in its pi orbital: $(tr)^3(pi)^1$. Sigma electrons form sigma bonds that link an atom to its neighbors and thus play an important role in determining the topological environment of the atom and, ultimately, the bond topology of the entire molecule.

Kier and Hall characterize an atom's topological environment by its delta value (δ), given by the number of sigma electrons or sigma bonds, although they discuss a number of essentially equivalent interpretations of δ. Because sigma electrons are generally less mobile than pi electrons, they tend to be less involved in interactions with other atoms, either within the molecule or in other molecules, than either pi or lone-pair electrons. An important index related to δ is the valence delta value (δ^v), which gives the count of sigma, pi, and lone-pair valence electrons (exclusive of bonded hydrogens): $\delta^v(\text{carbon}) = 4 - h$, $\delta^v(\text{nitrogen}) = 5 - h$, and $\delta^v(\text{oxygen}) = 6 - h$. Taking the difference

$\delta^v - \delta$ yields the number of pi and lone-pair electrons, a quantity shown by Kier and Hall to be related to an atom's valence-state electronegativity. With a suitable modification to treat bonded hydrogens, an I-State of an atom is essentially given by the ratio of the count of pi and lone-pair electrons, $\delta^v - \delta$, to the count of sigma bonds, δ. The latter provides a measure of the number of possible modes of "through-bond" interactions open to the atom.

To determine an atom's E-State once its local environment has been characterized by its I-State, it is necessary to incorporate the influence of the remainder of the molecular environment. Kier and Hall do this in a simple but intuitively appealing way as a sum of pairwise "atom–atom" interactions, each being made up of two factors. The first factor represents the perturbation of, for example, the ith atom's I-State, I_i, by the I-State of one of the remaining atoms in the molecule, for example, I_j, and is given explicitly by the difference in I-State values of the two atoms, $I_i - I_j$. The second factor is related to the square of the "distance" between the two atoms, r_{ij}. Since molecules are generally represented by their chemical graphs in this work, a distance measure related to graph-theoretical distance, such as the number of bonds on the shortest path between two atoms, should be employed. Here, the number of atoms (number of bonds + 1) on the shortest path is used, although a generalization to physical distances is also presented. The form of the perturbation term is taken as the ratio of these two factors, $\Delta I_{ij} = (I_i - I_j)/r_{ij}^2$, where it is clear that the distance factor attenuates the perturbation due to the I-State difference; i.e., more distant pairs of atoms experience smaller perturbations than the same pairs located in closer proximity.

Since I-State values and distances are always positive, ΔI_{ij} is positive if $I_i > I_j$ and negative otherwise and thus, $\Delta I_{ij} = -\Delta I_{ji}$. Hence, the influence of the I-State of the jth atom on the I-State of the ith atom is opposite to that encountered in the reverse direction. Kier and Hall interpret the influence of the surrounding atoms in a manner reminiscent of electronegativity equalization, where electrons are seen to flow toward the most electronegative atoms (i.e., along a positive "electronegativity gradient"). Here, the sign of ΔI_{ij} determines the "direction of influence" of the I-States of surrounding atoms.

The E-State of an atom is then explicitly given by the sum of its I-State value and the values of all the perturbing terms due to the remaining atoms in the molecule. Mathematically, the E-State for the ith atom is given by the expression $S_i = I_i + \Sigma_j \Delta I_{ij}$, where the sum is over all other atoms in the molecule. Interestingly, the sum of a molecule's E-State values is equal to the sum of its I-State values, i.e., $\Sigma_i S_i = \Sigma_i I_i$, and thus the sum depends only on the number and type of atoms in the molecule and not on their mutual interactions.

Initially, Kier and Hall treated only first-row atoms, such as carbon, nitrogen, and oxygen. In their later work, they developed suitable modifications of their basic I-State expression to handle hydrogen as well as second and higher row atoms. Thus, E-States for atoms in most molecular environments of interest in drug design are now available. Within this general framework, calculation of electrotopological indices is straightforward. The main issue is then whether these indices encode information on molecular features that is useful in elucidating structure–activity and structure–property relationships. As is true with most new molecular indices, it has taken time to clarify their "meaning" and to develop appropriate extensions to allow the treatment of, for example, polar hydrogens (HE-State indices) and higher row atoms such as chlorine and bromine.

Not surprisingly, earlier papers by Kier and Hall dealt with the nature of E-State indices and their extensions, along with several quantitative structure–activity relationship (QSAR) applications. However, since approximately 1995, the scope of applications has grown and includes studies in electrophoretic mobility, gas-chromatographic retention indices, neural networks, solubility, chemical reactivity, and an extension of the concept of free valence. In these applications, E-State indices are usually used in conjunction with other typical types of QSAR indices, although a few of the applications deal exclusively with E-States. Since the initial paper by Kier and Hall in 1990, more than 40 papers on the analysis and application of E-State indices have been published. While most of the applications have a strong QSAR flavor, some have moved into new and interesting directions, such as E-State molecular fields and molecular similarity.

With the explosive growth in comparative molecular field analysis (CoMFA)-based 3-D QSAR studies, E-State molecular fields provide another alternative that can be correlated with bioactivity. In fact, a recent 3-D QSAR study of the classic CoMFA steroid test data set showed that E-State/HE-state fields produced better correlations with experimental bioactivity of these compounds than did the more traditional steric, electrostatic, and hydropathic fields typically used in CoMFA studies.

Because E-State indices provide an atom-level characterization of molecules, they seem ideal as a basis for molecular similarity analysis. An initial study showed that "atom-type" E-States, obtained by summing the E-State values of all atoms of the same type within a molecule, provide molecular descriptors suitable for identifying similar compounds. This opens the door to database searching, which is becoming an important activity within the pharmaceutical industry, where very large compound databases exist. In the future, the amount of searchable structure and property information stored in electronic databases will increase substantially in all chemically related areas, so that improved descriptors for more effective and efficient similarity searches will be quite important. E-State indices may play a useful role here, too, but more research needs to be done to develop new similarity measures that will take full advantage of the level of atomic characterization provided by the E-State indices.

Now that the basics of E-States have been developed, extended, and tested in a number of applications, the real fun begins. It seems to me that both the methodology and the applications of E-State indices provide fertile ground for investigation. For example, Kier and Hall have briefly explored the use of geometric rather than the graph-theoretic distance measures used in the ΔI_{ij} perturbation terms, but more exploration is needed. Other procedures for computing E-State values should also be considered. For example, procedures that take account of the effect on S_i and S_j of atoms located on or near to the "bond path" connecting the ith and jth atoms should be investigated further; currently, only the atoms at the two "end points" of the path are considered. Kier and Hall have proposed a scheme for computing E-States iteratively that suggests one way of addressing this problem.

It is important, however, to remember the basic nature of E-States to avoid the kind of excessive tinkering that could destroy their elegant simplicity, which I believe is a strong point of the Kier and Hall approach. On the other hand, tinkering in the application area should be encouraged to ensure that the power of the E-State approach is realized fully. Many interesting classes of problems remain to be investigated. Of particular importance in drug design is the development of scoring functions for estimating ligand–protein and protein–protein binding free energies. Although considerable work is being done in this area, a totally satisfactory solution to the problem has not been achieved. Perhaps an extension of the bond E-States developed by Kier and Hall to include hydrogen and other noncovalent bonds would provide a suitable framework for a new scoring-function procedure. New measures of molecular similarity that go beyond atom-type E-States and treat all the E-States of a molecule might lead to more powerful similarity methods. Finding new ways of exploiting E-State fields should also be of value, especially since preliminary 3-D QSAR studies using such fields yielded encouraging results.

The use of E-State indices is in its infancy. Nevertheless, based on the type and variety of studies carried out to date, I believe that they have a promising future. This book provides extensive coverage of the subject, presented in a didactic manner that will assist those unfamiliar with the subject and that will provide new insights to those on more familiar ground. It should be on the bookshelf of every scientist who is seriously pursuing studies of the relationship of a molecule's structure to its chemical, biological, and physicochemical properties.

Gerald M. Maggiora
Kalamazoo, Michigan
June 1998

PREFACE

Combinatorial chemistry has burst onto the drug design scene like a tornado, making available huge numbers of compounds for lead discovery. Prior to this innovation, compounds were synthesized one at a time and data sets for quantitative structure–activity analyses (QSAR) were counted in the dozens at most. It is now possible to prepare and test (with high-throughput screening) hundreds or even thousands of compounds in a very short time. This vast amount of information is nearly useless without a systematic and relatively simple way of identifying and categorizing the molecules synthesized and tested.

The development of useful structure–activity models and their exploitation are the essential sequel to combinatorial chemistry. These models are necessary for the rational prediction of candidate compounds for more detailed analyses. The generation of this vital information depends upon the availability of methods for representing structure at both the molecular and the atoms-in-molecule level. These methods must be nonempirical, coherent, easy to use at high speed, and directly interpretable in terms of principles of chemical structure. Such information is also of great importance for the design and evaluation of new combinatorial libraries, both real and virtual. The electrotopological state (E-State) index is an ideal candidate for the applications needed to maximize the value of combinatorial chemistry and to create QSAR models.

It should be recognized that the E-State methods described herein are applicable to modeling other biological properties such as toxicity and bioaccumulation, as well as physicochemical properties, including aqueous solubility and chromatographic retention indices. An enormous benefit of this broadly based and widely applicable paradigm is the opportunity to investigate a range of property modeling problems while operating in the same structure representation genre. It is not necessary to shift from one formalism to another across the range of problems facing the chemist who is designing molecules with specified properties, whether in the drug-design arena, the sphere of environmental problems, areas of odor, flavor, and taste, development of polymers, or any area in which noncovalent interactions dominate the property of interest.

The E-State index was developed a decade ago to satisfy the need for an atom- or group-centered numerical code to represent molecular structure for use in structure–activity analyses. At that time, the primary focus was to illuminate specific parts of a molecule for use in this modeling activity. A series of papers by the authors between 1990 and 1992 and a doctoral dissertation by Nikhil Joshi at Virginia Commonwealth University in 1993 revealed the method and its value in these applications. About this same time, the burgeoning of combinatorial chemistry became prominent along with the need for database management, including applications of the concepts of similarity, diversity, and searching. These needs triggered our keen interest in the ability to label or code parts of molecules in order to design and use them in huge collections of compounds for molecular design.

The two central ideas in the E-State concept are relatively new: (1) representation of an atom in a molecule in terms of an intrinsic state perturbed by all other atoms in the molecule and (2) an atom intrinsic state defined as the ratio of valence state electronegativity to the number of skeletal bonds or avenues over which electron density is distributed. We believe that this latter concept is a basic characterization of an atom in a covalent molecule, encoding potential for intermolecular interaction. Likewise, the electrotopological state encodes basic structure information, suggesting broad applicability based on molecular structure rather than on any particular mechanism.

It has become evident that the need for an internally consistent scheme of atom and group coding is important in these procedures. The E-State index is a logical candidate for these applications; E-State indices encode significant structure information in a highly usable form that can be rapidly computed. The creation of this basically new structure representation has led us to assemble a comprehensive treatment of the development, significance, extensions, and applications of the E-State method under one cover. The features of the E-State formalism presented here are very much in accord with the ideas expressed by Milne (1997), who has called for more emphasis on heteroatom encoding in topological indices. He has also raised the challenge to devote more energy to the use of structure–activity equation models to address the inverse problem, the predictions of candidate structures. We feel that the E-State formalism lends itself to just such an endeavor and we are hard at work on this engaging problem.

Our strategy in constructing this book has been to blend the elements of a textbook, a review, and a research monograph under one cover. As a textbook, the systematic development of the E-State formalism is followed by examples. Questions and problems for the reader to solve punctuate several chapters. As a review, a number of studies from the literature are described in order to illustrate the general versatility and breadth of the paradigm. As a research monograph, we have introduced a number of unpublished variations and extensions of the basic E-State concept. These studies include a treatment of onium groups, a bond E-State index, a fragment-based database search technique, a polarity index, and some primitive ideas about a 3-dimensional E-State method. We anticipate that some of these studies will evolve into fully developed publications, but they are shared with the reader here in order to stimulate possible participation in their progress.

This book is assembled as a series of chapters plus a relatively new technique of including a compact disc to facilitate calculations on molecular structures. There are exercises scattered through the book to enhance the understanding of the method and applications. The first chapter discusses the concept of structure and describes the general framework in which the E-State method resides. Chapter 2 reveals the development and derivation of the E-State formalism with consideration of the influence of structure change on these values. Some exercises for the reader provide practical experience, hopefully for later use. Chapter 3 reveals the significance and interpretation of the method including some comparisons with other atom-level parameters and with the concept of free valence.

Chapter 4 describes some enhancements and extensions of the E-State method. Three important advances in the E-State method are shown. The first is the introduction of a parallel formalism for hydrogen atoms. The hydrogen E-State values permit encoding of polar hydrogen atoms as well as those that are nonpolar or lipophilic in character. The second is the development of an atom-type index that has significant possibilities in database management and property prediction for large heterogeneous data sets. The third is the use of the E-State as a field in schemes such as the comparative molecular field analysis (CoMFA). Strategies for the use of E-State indices are discussed in Chapter 5. Current mathematical methods including neural networks, partial least squares, and orthogonalization of a set of indices are shown to be applicable with E-State studies. Some attention to experimental design is included in this chapter. The use of the E-State in database applications is described in Chapter 6. Schemes are presented to show how the E-State indices can be used in database organization and searching. A basis for molecular similarity is also given for the E-State.

Chapters 7, 8, and 9 give examples of the use of the atom-level E-State in structure–activity analyses (using topological superposition), the use of E-State indices with other indices, and the use of the atom-type indices, respectively. Some exercises for the reader are included. The final chapter is a look-ahead passage where the authors reveal future possibilities for the E-State method together with possible changes and improvements.

No book is written entirely by the authors. There are many people who contribute, if only as critics or enthusiasts. Others type, draw, or provide technical assistance. We recognize these people and give them our heartfelt thanks. Martha Kier and Camille Vann helped with typing. Dorla Hall read the manuscript for final proofreading. Matthew Dowd rendered some of the figures and drawings. Our most recent collaborators include Glen Kellogg, Carolina de Gregorio, Nikhil Joshi, Cynthia Shiek, Bryan Grella, James Burnett, Jonathan Gough, David Dunmire, and Timothy Vaughn. We give special appreciation to Jack Frazer who early recognized the significance and potential of the E-State. We appreciate the serious reading given by Gerald Maggiora and his preparation of the foreword to this book. Further, we appreciate the comments provided by two readers of the manuscript. Douglas Kitchen provided insight from an overall perspective. Special acknowledgment is given to Michael Tute, who gave us thoughtful and detailed commentary on each chapter. The staff at Academic Press, including David Packer and Linda Klinger, are acknowledged and thanked for their help and input. We give special thanks to Joanna Dinsmore, who gave careful and thoughtful attention to copy editing and production. We thank G&S Typesetters for the quality production of the actual text. Finally, we acknowledge the encouragement and support of our colleagues at our respective institutions, Virginia Commonwealth University and Eastern Nazarene College.

REFERENCE

Milne, G. W. A. (1997). Mathematics as a basis for chemistry. *J. Chem. Inf. Comput. Sci.* 37, 639–644.

Lemont B. Kier, Richmond, Virginia
Lowell H. Hall, Quincy, Massachusetts

CHAPTER 1

Introduction

I. THE GENERAL PROBLEM

The development of new drugs with pharmacological efficacy and clinical utility is now a major activity in the chemical and biological sciences. From successes in midcentury, there has evolved a sense of confidence in the ability to make some predictions in the design of molecules with desired characteristics. Today, we are on the threshold of rational drug design based on the occasional ability to recreate and model the molecular-level scene of action of a ligand molecule and an effector using computer graphic simulations. From such a model, variations in structure may be made with the objective of improving the encounter possibilities leading to a better binding or reaction between the two. Wherever the effector and whatever the mechanism to produce an effect, there is the possibility of designing drugs from these models.

The opportunity to use this approach in drug design is limited to situations in which there is considerable information, such as the specific sites of action or the structure of the enzymes involved and the ligands that interact. In the absence of a well-defined molecular target, it is necessary to adopt other strategies to develop a drug. Beyond the attribute of biological activity, in the realm of pharmaceutical properties, such as absorption, solubility, distribution, and excretion, it is not possible to build a model around a macromolecular system with an active site. These properties are in the domain of molecular systems in which something resembling an effector, if this is even an appropriate term, is an evanescent group of water molecules in intimate contact with solutes or protein surface fragments forming a complex system. It would be most useful indeed to have a method or series of methods, all based on the same approach to molecular structure, to model all these important types of properties and activities.

This reflection leads us to consider an alternative approach to drug design—one that we call quantitative information analysis (Kier and Hall, 1997). In this approach, we present a series of molecules to a biological system and measure the properties of interest (Fig. 1.1). We are concerned with the structure of the ligand (A) and the numerical readout from the system (B). The excluded middle is a hierarchy of complex systems, each with emergent properties ultimately fueling the creation of still higher orders of complex systems.

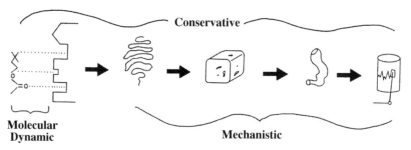

FIGURE 1.1   In a biological system, the relation between the dynamic molecular level and the various higher level systems which ultimately lead to a measured signal which may be used in SAR analysis is shown.

Ultimately, a system yields the emergent properties that we measure (B) which are clinically useful. The essence of this approach to drug design is the creation of a model relating some attribute of the molecules introduced to the biological system and the measured properties emerging from a remote consequence of encounter. We make the general statement that $B = f(A)$. We do not expect to acquire much wisdom in the form of mechanistic models of many of the intervening events. This is the realm of molecular biology. Our objective is to modify the molecular structures to improve the measured output in a conservative model. We call this general approach quantitative information analysis. It is not a new paradigm, but we have used this generic term to describe attempts to quantitate the drug design process and to signal a broader point of view of the overall design process.

The process of quantitative information analysis has followed two paths over the years. One of these paths has led to the use of physical properties to describe a molecule in a model, relating it to a measured response (Hansch, 1969). This approach has come to be called quantitative structure–activity relationship (QSAR), where the word structure was loosely used to denote measured or estimated physical properties. At the same time, an alternative paradigm to physical property models arose in the form of theoretical models of structure derived from molecular orbital theory (Kier, 1971). These models attempted to relate defined structure quantitation to measured responses in an effort to create a predictive model. Since that time, innovations to the concept of structure have emerged that extend the practical utility of QSAR models. In particular, the introduction of graph-based topological indices two decades ago has made possible a description of structure that is simple and demonstrably valuable in predictive power (Randic, 1975; Kier and Hall, 1976, 1986). This path to molecular description is the focus of this book.

It is important at the outset that we distinguish between physical properties and structure. To some, this may seem obvious and trivial. However, this early assumption (that property is equated with structure) has had an impact on many scientists which is so pervasive that the distinction has become widely blurred. We devote space to the clarification of the distinction before we proceed.

II. WHAT IS STRUCTURE?

The answer to this question is not as obvious as it may seem, in view of the misuse of the word in journal articles dating back three decades. Attempts to rectify these misstatements were made by Norrington *et al.* (1975), who proposed the term property activity relationships as a distinctive subclass of structure–activity models. The authors have included brief descriptive statements in many articles using topological structure indices during the past 20 years (Hall and Kier, 1991, 1993; Hall *et al.*, 1991, 1993; Kier *et al.*, 1993; Kier and Hall, 1994). Recently, Testa and Kier (1991) and Testa *et al.* (1997) presented detailed descriptions of these attributes in a comprehensive review. Hoffmann and Laszlo (1991) have written eloquently on this subject from a broad chemical perspective. For our purpose in this book, to present information about a new subclass of molecular descriptors, we offer the following brief essay on molecular structure and its representation.

Molecular structure is a collective term for codes by which we describe a molecule in quantitative terms. It is a model of the form of a molecule, which produces a series of functions called properties. The functions or properties are measured attributes which, in the case of molecules, are averaged values of responses to chemical or physical input into a system containing the molecule in question. The numerical values of the properties present a mosaic of information about the system. From this information, we weave a model of the structure which may have given birth to our measurements. Structure is a model; it is a guess as to what is there, functioning in response to our physical inquiry. A representation of what is there is a model of the "form" from which understanding and prediction may be possible and practical. In this sense, any representation may be incomplete but judged adequate for a particular purpose. The function or properties of a molecule are dependent on the form or structure. It is an immutable relationship: Structure is the antecedent to properties and form precedes functions. Figure 1.2 captures these relationships. How do we encode structure? Testa and Kier (1991) and Testa *et al.* (1997), in describing a hierarchy of indices encoding properties of increasingly complex systems, have addressed this question as shown in Table 1.1.

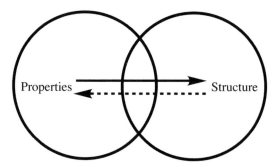

FIGURE 1.2   A diagram of the relation between structure and properties which indicates that structure is the antecedent to property.

A. THE CHEMICAL GRAPH

One way to describe a molecule is to begin with an image called a graph, a network reflecting the content and adjacency relationships of the ingredients. We consider a molecule as a collection of atoms together with a set of relationships based on the network of chemical bonds. The collective information from this depiction is the topology of the molecule. In familiar terms, we are looking at the skeleton of a molecule. Benzene is invariably drawn as a simple hexagon with a circle in the middle. Each vertex in the drawing represents a carbon atom with implied attached hydrogen atom(s), whereas each line represents a chemical bond between two neighboring carbon atoms. The circle represents a bond

TABLE 1.1   Levels of Structure and Properties

Level	Property/attribute	Structure
Elementary	Molecular weight Ionization potential	Atom count and type
Geometric	Molar volume Surface area Flexibility Conformation	Atom connections Bond orders and topology
Intermolecular interaction	Boiling point Solvation Partitioning	Atom type and connectivity
Biological interaction	Biological properties	Molecular topology and electron distribution

of a different kind, existing uninterrupted over all atoms in the molecule. The hydrogen atoms are omitted for simplification, and our chemical intuition tells us that they are there, satisfying valence requirements of the carbon atoms. The letters depicting the different elements represented in the graph encode the information of electron counts partitioned according to the atom type among sigma, pi, and lone-pair orbitals.

From these chemical graphs, there has evolved during the past 20 years a new paradigm of molecular structure description called molecular connectivity (Randic, 1975; Kier and Hall, 1976, 1986). From this concept, a series of indices have become widely used in QSAR studies in which the predictive power of relating equation models has been sought to guide rational molecular design. Molecular connectivity indices are whole-molecule codes, quantifying structure attributes associated with the entire molecule. It is possible to dissect submolecular contributions to properties through comparisons among related series of molecules. The information, however, is still based on the structure of the molecule.

B. STRUCTURE AND SUBSTRUCTURE

Equation models expressing the effect of molecular change on properties are of great value in molecular design. Two decades of exploitation have revealed the intrinsic value of these models. There is, however, an alternative approach to the encoding of molecular structure which may have great value in molecular design—the quantitation of structure at the level of the atom or group. It is increasingly clear that drug–receptor, enzyme–substrate, and binding interactions are processes that invoke prominent roles for specific atoms or groups within a molecule—that is, the concept of the pharmacophore in a ligand and the active site of an effector. If a molecular approach is taken in the development of a quantitative model, the identity and pattern of variation of these specific submolecular features may be obscured. This loss of information can happen whether the model is derived from physical properties as parameters or is made up of structure indices from whatever paradigm. There is a need to encode and quantify molecular substructure and to include this concisely in various approaches in order to understand more completely the effect of molecular change on biological activity. Figure 1.3 depicts the four realms of structure description within the realms of property–structure space and whole molecule–fragment space. It is this perceived need for an atom-level index that has fueled our creative efforts. These efforts, the results, their use strategy, and the applications in drug research are the topics of this book.

Information Domains of QSAR Parameters

	Physical Properties	Structure
	Measurement:	Calculated (modeled):
Whole Molecule	log P, R_M reaction rates	M.O. Energies Molecular Connectivity Kappa shape Information Theory
	Measurement:	Calculated (modeled):
Atom or Fragment	x-ray analysis from additivity schemes: Hansch pi Hammett sigma Taft E	M.O. charges M.E.P. maps Atom-centered graph theory indices

FIGURE 1.3   A depiction of the four realms of structure description within property-structure space.

III. ASPECTS OF ATOM-LEVEL DESCRIPTION

There are three aspects of an atom-level description that bear attention if a broadly useful quantitation is to be achieved. The first of these arises as a consequence of the traditional dissection of atom-level or fragment-level structure into two attributes, the electronic and the spatial or topological. The second aspect which demands attention is the reality that an atom or fragment-level description must possess the information that the feature being encoded is a part of the whole (the molecule). Each atom or group is influenced by all parts of the molecule. The generality of a code for each substructure feature cannot accurately transcend the confines of the molecule from which it is derived. Finally, the issue of the relationship of the structure code of the fragment to the three-dimensional character of the molecule must be addressed in order to construct a generally useful index. We shall consider each of these issues in turn.

A. THE ELECTRONIC/STERIC DUALITY
OF STRUCTURE DESCRIPTION

It has become a common practice to employ a reductionistic approach to describing molecular or fragment structure. The molecule or fragment is dissected into attributes that take the form of the duality, electronic and steric structure. The electronic component is derived from an estimate of the local charge density, often a population analysis from a semiempirical molecular orbital calculation. The importance of this electronic description was demonstrated early, when charge densities were employed as reaction indices useful in studying chemical and biological phenomena (Kier, 1971; Pullman, 1963; Streitwieser, 1961; Richards, 1977).

The second attribute of this duality is an estimate of what is called steric, spatial, or topological structure, although these terms are not rigorously equivalent. The approach here has usually been based on a geometric estimate of some limit of the probability domains of valence electron orbitals. The van der Waals radius is commonly employed. An early example of this separation of attributes is the Hansch approach based on the assumed isolated contributions of electronic, steric, and lipophilic characteristics of a molecule or fragment (Hansch, 1969). A recent example is the use of assumed orthogonal fields arising from electrostatic interactions and steric effects in the comparative molecular field affects analysis (Cramer *et al.*, 1988).

The separation of these contributions has been of value in analyzing the structure influences on a property. It must be realized, however, that electronic and steric attributes are manifestations of the same structure attribution—the probability distribution of electrons in a molecule. It has been the common practice to separate these effects into phenomena more familiar to us than is the case in the true state. Electronic structure is likened to a density or a probability distribution, whereas a steric description coincides with the awareness of boundaries and solid encounters in the macroscopic world. It should be understood, however, that these two attributes arise from the same atom, fragment, and molecular structure: the presence and probable positions of valence electrons on the mantles of atoms, fragments, and molecules.

B. THE LOCALIZATION
OF STRUCTURE DESCRIPTION

The second aspect of atom and substructure description that we shall consider is localization of some of the estimates of charge or steric structure. This is not a problem in molecular orbital calculated attributes but molecular mechanics estimates bear this burden. Both steric structure and charge distribution are attributes dependent on neighbors near and remote in the molecule. The structure context of each atom or fragment is thus of prime concern in any attempt to quantitate structure. The atom-level code must not extract the atom from its molecular network but represent it fully within its bonded and topological context.

C. THE DIMENSIONALITY QUESTION

A molecule is a complex entity which has extension in three-dimensional space. This is implicit in stereochemical models. Some information is lost when a molecule is portrayed in two-dimensions on a slide or printed page. The two-dimensional depiction, however, encodes all the connectivity of a molecule. The interatom relational information implied in a graph is less than that implicit in a three-dimensional depiction. Attributes such as *cis* and *trans* isomerism are not conveniently extracted from a graph. The retention of the connectedness of atoms and the presence of neighbors of various degrees, however, are retained in the graph description of a molecule. The graph of a molecule may be thought of as accurately representing a completely networked system in which there is a real probability of inductive influence everywhere in the molecule. This influence is greatest among bonded neighbors but it may extend well beyond this propinquity in certain circumstances. The ebb and flow of these influences are what impart to each atom (or group) in a molecule its discreteness and its uniqueness relative to all other like atoms in molecules with different structures.

A growing body of evidence from practical experiences with databases reveals the practical value of topological and graph-based representations, sometimes loosely called two-dimensional structure representations (Cummins *et al.*, 1996; Cheng *et al.*, 1996; Lewis *et al.*, 1997; Brown and Martin, 1998; Zheng *et al.*, 1998a,b). Certainly, the ease of acquisition, cost, and relative significance of two-dimensional structure information, when considering huge databases, is a compelling reason to exploit it in drug design. The necessity of processing data from combinatorial chemistry requires these attributes. Three-dimensional methods seeking conformational definition have varying degrees of uncertainty; the structure portrayed by a graph has no uncertainty in the information that we use. The use of a graph to define and to process structure information is a viable framework for practical applications in drug design. Methods based on this model should be the first in line for exploitation in database management or structure–activity modeling.

It should also be noted that very significant information on three-dimensional structure is implicit in the set of connections contained in the chemical (molecular) graph. A graph containing only — CH_2— groups is clearly a cyclic alkane; if the graph also has two — CH_3 groups, then it corresponds to a normal alkane. If the graph contains >CH— groups as well, the molecule is branched. If quaternary carbons, >C<, are present, especially if they are adjacent, there is significant steric crowding. If there are alternating double bonds, there is significant planarity. These illustrations indicate that the molecular graph does encode (implicitly) significant three-dimensional (3-D) or topographic information. Although certain 3-D features are not implicit in the graph formalism, it is not accurate to refer to topological indices as simply 2-D. The dimensional adjective is inappropriate here. The richness and depth of structure information plus the effectiveness of topologically based QSAR models are more useful accolades.

IV. SUMMARY

The need for atom- and fragment-level structure descriptors is evident in the face of the explosion of compound production from combinatorial chemistry. This lead-generating method must be followed by a systematic approach to the definition of a pharmacophore and a modeling of the structure–activity relationship. One way to accomplish this is through the use of nonempirical descriptors based on the chemical graph. Furthermore, the necessity to model aqueous solubility and other physicochemical properties in addition to toxicity requires a broadly based structure representation which is rapidly computed and readily interpreted in terms of molecular structure. These objectives are met through the use of a series of indices to be described and analyzed in this book.

V. A LOOK AHEAD

In Chapter 2, we describe a novel method for encoding the electronic structure, the topology, and the interrelation of an atom in a molecule to all other atoms in that molecule. This method produces a set of nonempirical parameters for each atom or hydride fragment in a molecule based on the chemical graph.

REFERENCES

Brown, R. D., and Martin, Y. C. (1998). An evaluation of structural descriptors and clustering methods for use in diversity selection. *SAR QSAR Environ. Res.* 8, 23–40.

Cheng, C., Maggiora, G., Lajiness, M., and Johnson, M. (1996). Four association constants for relating molecular similarity measures. *J. Chem. Inf. Comput. Sci.* **36**, 909–915.

Cramer, R. D., Patterson, D. E., and Bunce, J. D. (1988). Comparative molecular field analysis (CoMFA). Effect of shape on binding of steroids to carrier proteins. *J. Am. Chem. Soc.* **110**, 5959–5967.

Cummins, D. J., Andrews, C. W., Bentley, J. A., and Cory, M. (1996). Molecular diversity and chemical databases: Comparison of medicinal chemistry knowledge bases and databases of commercially available compounds. *J. Chem. Inf. Comput. Sci.* **36**, 750–763.

Hall, L. H., and Kier, L. B. (1991). The molecular connectivity chi indices and kappa shape indices in structure–property relations. In *Reviews of Computational Chemistry* (D. Boyd and K. Lipkowitz, Eds.), pp. 367–422. VCH, New York.

Hall, L. H., and Kier, L. B. (1993). Design of molecules from quantitative structure–activity relationship models. III. Role of higher order path counts: Path Three. *J. Chem. Inf. Comput. Sci.* **33**, 598–603.

Hall, L. H., Mohney, B. K., and Kier, L. B. (1991). The electrotopological state: An atom index for QSAR. *Quant. Structure–Activity Relationships* **10**, 43–51.

Hall, L. H., Kier, L. B., and Frazer, J. W. (1993). Design of molecules from quantitative structure–activity relationship models. II. Derivation and proof of information transfer relating equations. *J. Chem. Inf. Comput. Sci.* **33**, 148–152.

Hansch, C. (1969). A quantitative approach to biochemical structure–activity relationships. *Acc. Chem. Res.* **2**, 232–239.

Hoffmann, R., and Laszlo, P. (1991). Chemical structures: The language of the chemist. *Angew. Chem.* **103**, 1–16.

Kier, L. B. (1971). *Molecular Orbital Theory in Drug Research.* Academic Press, New York.

Kier, L. B., and Hall, L. H. (1976). *Molecular Connectivity in Chemistry and Drug Research.* Academic Press, New York.

Kier, L. B., and Hall, L. H. (1986). *Molecular Connectivity in Structure–Activity Analysis.* Wiley, London.

Kier, L. B., and Hall, L. H. (1994). The generation of molecular structures for a graph based equation. *Quant. Structure–Activity Relationships* **12**, 383–388.

Kier, L. B., and Hall, L. H. (1997). Quantitative information analysis: The new center of gravity in medicinal chemistry. *Med. Chem. Res.* **7**, 335–339.

Kier, L. B., Hall, L. H., and Frazer, J. W. (1993). Design of molecules from quantitative structure–activity relationship models. I. Information transfer between path and vertex degree counts. *J. Chem. Inf. Comput. Sci.* **33**, 143–147.

Lewis, R. A., Mason, J. S., and McLay, I. M. (1997). Similarity measures for rational set selection and analysis of combinatorial libraries: The diverse property-derived (DPD) approach. *J. Chem. Inf. Comput. Sci.* **37**, 599–614.

Norrington, F. E., Hyde, R. M., Williams, S. G., and Wooten, R. (1975). Physicochemical activity relations in practice. 1. A rational and self-consistent data base. *J. Med. Chem.* **18**, 604–608.

Pullman, B. (1963). *Quantum Biochemistry.* Wiley, New York.

Randic, M. (1975). On characterization of molecular branching. *J. Am. Chem. Soc.* **97**, 6609–6615.

Richards, W. G. (1977). *Quantum Pharmacology.* Butterworths, London.

Streitwieser, A. (1961). *Molecular Orbital Theory for Organic Chemists,* pp. 329–341. Wiley, New York.

Testa, B., and Kier, L. B. (1991). The concept of structure in structure–activity relations by studies and drug design. *Med. Res. Rev.* **11**, 35–48.

Testa, B., Kier, L. B., and Carrupt, P.-A. (1997). A systems approach to molecular structure, intermolecular recognition, and emergence–dissolvence in medicinal research. *Med. Res. Rev.* **17**, 303–326.

Zheng, W., Cho, S. J., and Tropsha, A. (1998a). Rational combinatorial design. 1. Focus 2-D: A new approach to the design of targeted combinatorial chemical libraries. *J. Chem. Inf. Comput. Sci.* **38**, 251–258.

Zheng, W., Cho, S. J., and Tropsha, A. (1998b). Rational combinatorial design. 2. Rational design of targeted combinatorial peptide libraries using chemical similarity probe and inverse QSAR approaches. *J. Chem. Inf. Comput. Sci.* **38**, 259–268.

CHAPTER 2

The Electrotopological State

I. THE MOLECULE AS A COMPLEX SYSTEM

The study of chemical and physical phenomena that are connected with biology is a study of molecules—their form, function, and fluctuation (Testa *et al.*, 1997). The molecule is a complex system by any definition. It is a consequence of the coming together of atoms, each of which surrenders its identity and function in a process called dissolvence (Testa and Kier, 1997). The result is an emergent set of properties associated with the complex system that is a molecule. The classical reductionist approach to the study and characterization of a molecule is to dissect it into its antecedents, the atoms. This information is intended to epitomize the molecule and to lead to some understanding of the function recorded as physical, chemical, and biological properties. However, this is an insufficient process; lacking is some proper reconstitution of the parts to make a whole. A linear combination of the attributes of atoms will not foretell the functions of a molecule. This situation is encountered when assuming that the whole is equal to the sum of its parts. Some nonlinear process of combining atom information must be employed—a process of synthesis.

We recognize these aspects of molecular complexity as we seek some understanding and quantitation of the molecule as a whole and the important features of which it is composed. Furthermore, we aim to develop representations of molecular structure which are useful in modeling the relationship with biological activity and/or physical properties. As a result of sound experiments and good models, we are aware that fragments of molecules play significant roles in their biological function. It is important to identify these fragments, recognizing that they are a part of the whole and at the same time playing semi-independent roles in biological encounters. This challenge has led us to consider the development of indices encoding structure information about the fragments within the context of the entire molecule.

II. ATOM INFORMATION FIELDS

Any approach to the quantitation of atoms or fragments within a molecule must be built on the relationships and forces operating within the complex system of the molecule. We can view each atom in a molecule as existing in a field within a molecule in which all other atoms participate. This field participation is characteristic of any atom in a particular molecule. The methyl group in toluene (Scheme I) is different from the methyl group in acetic acid (Scheme II) by virtue of its context, despite its intrinsic state as a methyl group. Quantifying the methyl group requires both an identity as a methyl group and its modification through the relationships to all other atoms in the molecule in which it resides.

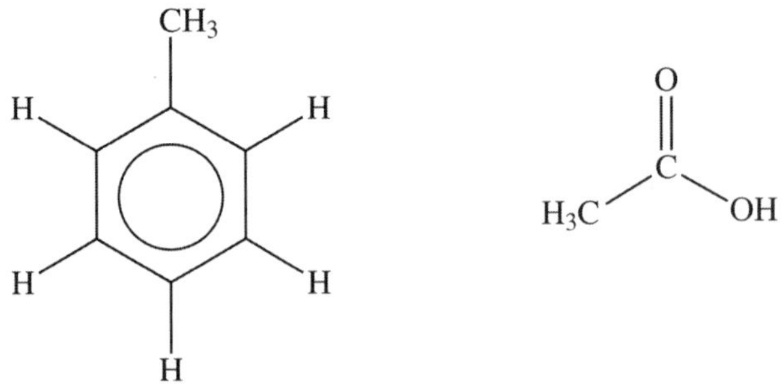

(I, II)

The influence of all other atoms in toluene makes the methyl group unique relative to methyl groups in all other molecules. This modifying influence is propagated primarily through the formal bonding relationships within the molecule. It is a very good assumption that the covalent bonding structure or an abbreviation of it, the chemical graph of toluene, is the framework over which this influence is propagated. Consider the molecule of toluene, drawn as a chemical graph (Scheme III). There are characteristics of the methyl group which must be identified in order to describe it in a quantitative way, unique to its presence in toluene. We consider each of these characteristics in turn.

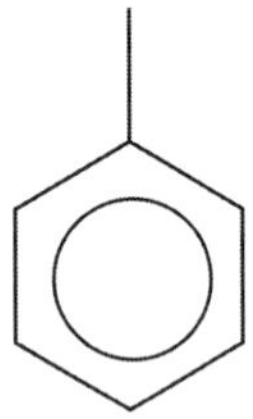

(III)

III. THE INTRINSIC STATE OF AN ATOM

Continuing our example of the methyl group in toluene, we recognize that all methyl groups in all molecules possess basic attributes that are identified as intrinsic. We can employ a single symbol to identify all methyl groups. The common and essential attributes constitute an intrinsic state for any methyl group. Note that we treat the methyl group as a single unit of four atoms since its integrity as a functioning system is well documented. This approach is consistent with the hydrogen-suppressed graph.

What are the common attributes of methyl groups that comprise its intrinsic state? We are led to choose those which we believe have an important influence on chemical, physical, and biological properties. Three attributes are immediately apparent. The first of these is the elemental content. In the case of the methyl group, this is a composite of carbon and hydrogen atoms. A single atom in a molecule is described as a particular element. The second attribute that is part of its intrinsic state is the electronic organization. We represent this aspect as the hybrid state or more simply the valence state of the atom or group in a molecule. This characterization includes the counts of sigma, pi, and lone-pair electrons comprising the valence electrons of the atom. In the case of a group such as methyl, we must also encode the number of hydrogens to distinguish it from the other hydrides of carbon as we do in other groups containing nitrogen and oxygen.

A third ingredient in an intrinsic state description is the degree of adjacency or more generally the local topological state of the atom or group. This attribute is important in defining the position of the atom relative to the topology of a molecule. For example, if we compare the starred carbon atoms in the three isomers of pentane (Schemes IV, V, and VI), it is apparent that the spatial domain or accessibility of each carbon is different. The starred methyl group in pentane (Scheme IV) is a mantle fragment residing on the periphery of the molecule; hence, it is easily accessible to interactions with neighboring molecules. In contrast, the starred methylene group in isopentane (Scheme V) is located within the structure of the molecule with somewhat diminished accessibility to an intermolecular contact. Finally, the central atom of neopentane (Scheme VI) is buried deep within the molecule. Its accessibility to any contact or intermolecular interaction is virtually nil.

(IV, V)

(VI)

Some of these attributes have been considered by other investigators seeking to quantify atom-level information. Hall and Kier (1977) employed molecular connectivity to quantify partial charges on carbon hydride groups in alkanes. Randic (1984) reported on the development of an atom identification number from summed atom values. Other approaches include torsion topological descriptors by Nilakantan *et al.* (1987), an electron-topologic approach by Dimoglo (1988), a topological electronic index by Kalizan *et al.* (1986), a vertex topologic index by Klopman and Raychandhury (1988), and the development of atomic contributions for physicochemical properties by Ghose and Crippen (1987).

IV. THE GRAPH REPRESENTATION OF A MOLECULE

Common among these approaches is the use of the chemical graph to represent a molecule. The graph is a set of objects (atoms) and a set of relationships (bonds, connections, adjacencies, or information) among the objects. The bonding scheme of atoms is depicted as a network using lines to represent bonds and vertices to represent atoms. It is common to omit hydrogen atoms from carbon vertices since these are implied by valence rules. The result is called a hydrogen-suppressed graph. Hydrogen atoms associated with hetero-atoms are retained in the scheme for clarity. Multiple bonds are explicitly represented. Aromatic rings include pi bonds using canonical representation or a circle denoting the pi orbital annulus or as Kekule structures. Toluene (Scheme III) and acetic acid (Scheme VII) are shown as graphs according to this convention.

$$
\begin{array}{c}
O \\
\parallel \\
C \\
\diagup \quad \diagdown \\
\qquad OH
\end{array}
\qquad (VII)
$$

V. THE ATOM REPRESENTATION

A. THE DELTA VALUES

The depiction of an atom with its hybrid or valence state presents both a challenge and an opportunity. The challenge is to derive a meaningful representation that will carry valuable information in reflecting field effects. The oppor-

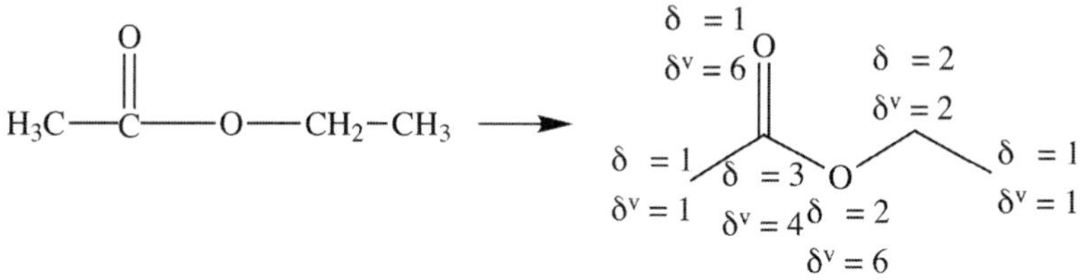

FIGURE 2.1   Ethyl acetate shown with the conventional structure formula and as a hydrogen-suppressed graph with simple and valence delta values for each vertex (atom/group).

tunity lies in the possibility of encoding not only the atom hybrid state but also the relative electronegativity of the atom in the molecule. The atom description is essentially a statement of electronic structure and the distribution of valence electrons among various orbitals in hybrid states. This is the defining characteristic among covalently bound atoms in organic molecules, including those of interest in biological systems. Using the chemical graph, we have encoded this information from counts of electrons (Kier and Hall, 1976, 1986; Hall and Kier, 1978). At this level of description, we refer to these quantities as graph primitives. It is from this basic information that we synthesize the higher level indices which encode significant structure information that becomes the basis for molecular representation.

We begin by writing the conventional chemical structure of a molecule such as ethyl acetate (Fig. 2.1) and then delete hydrogens associated with carbon atoms to form the chemical graph. We now associate with each atom two values used in our earlier work on molecular connectivity descriptions. The first is the simple delta (δ) value, which is the count of adjacent atoms other than hydrogen. In Fig. 2.1, the delta values in the molecule of ethyl acetate are shown. Note that just the adjacency is encoded. This is equivalent to describing just the sigma bond skeleton structure. A second set of delta values has been included for each atom in the molecule in Fig. 2.1. These are based on the total number of valence electrons on the atom minus those bonding hydrogen. These have been called valence delta values (δ^v) in earlier work. These two sets of delta values are the basic ingredients for the definition of the intrinsic state of atoms in molecules.

B. THE ELECTRONIC INFORMATION IN THE DELTA VALUES

The pair of delta values previously discussed provides a rich resource of information about atoms in molecules in our quest for a structure description. The information is presented from more than one perspective, adding insight to their significance. The simple delta value is defined as follows:

$$\delta = \sigma - h$$

in which σ is the count of electrons in sigma orbitals and h is the count of bonded hydrogen atoms.

The simple delta value, δ, encodes

1. The count of adjacent atoms excluding hydrogen
2. The count of sigma electrons on an atom excluding hydrogen
3. The count of bonds joining to an atom other than hydrogen
4. The immediate topological environment of the atom in the molecule

The valence delta value is defined as follows:

$$\delta^v = Z^v - h = \sigma + \pi + n - h$$

where π is the number of electrons in pi orbitals and n is the number of electrons in lone-pair orbitals.

The valence delta value, δ^v, encodes

1. The count of valence electrons on an atom other than to hydrogen
2. The count of sigma, pi, and lone-pair electrons excluding bonds to hydrogen

Table 2.1 summarizes the δ and δ^v values for carbon, nitrogen, oxygen, and fluorine atoms in a molecule. It is evident that $\delta^v - \delta = \pi + n$, the count of pi and lone-pair electrons on an atom in a molecule (Kier and Hall, 1981). This information provides a quantitative measure of the potential of the atom for intermolecular interaction and reaction. In addition, it has a high correlation with the Mulliken–Jaffe electronegativity of atoms in their valence states

TABLE 2.1   Delta Values for Carbon, Nitrogen, and Oxygen

Atom	Hybrid state	δ^v	δ
>C<	sp^3	4	4
=C<	sp^2	4	3
≡C—	sp	4	2
>N—	sp^3	5	3
=N—	sp^2	5	2
≡N	sp	5	1
—O—	sp^3	6	2
=O	sp^2	6	1
—F	sp^3	7	1

TABLE 2.2   Kier–Hall Electronegativity Values

Atom	$\delta^v - \delta$	$X_{KH} = (\delta^v - \delta) / n^2$ X_{KH}	X_M (eV)
=O	5	1.25	17.07
—O—	4	1.00	15.25
≡N	4	1.00	15.68
=N—	3	0.75	12.87
>N—	2	0.50	11.54
≡C—	2	0.50	10.39
=C<	1	0.25	8.79
>C<	0	0.00	7.98
—Cl	6	0.67	11.54
=S	5	0.55	10.88
—S—	4	0.44	10.14

Note. Abbreviations used: n, principal quantum number; X_{KH}, Kier–Hall electronegativities; X_M, Mulliken–Jaffe electronegativities expressed in electron volts.

(Mulliken, 1934; Hinze and Jaffe, 1962). These values, referred to as the Kier–Hall electronegativity (Kier and Hall, 1981), are shown in Table 2.2. The electronegativity of an atom in a molecule is of major importance within the context of the general information field described earlier. As a consequence, this simple statement of structure has a significant role in encoding the intrinsic state of the atom.

VI. THE INTRINSIC STATE ALGORITHM

The possibility of encoding a close approximation of the valence state electronegativity with such a simple index is of great value. Table 2.1 lists the values of δ^v and δ for several covalently bound atoms in their valence states. The close correlation between the Kier–Hall and the Mulliken–Jaffe electronegativity values is evident from Table 2.2. The derivation of an intrinsic state index labeled I begins with the use of the $\delta^v - \delta$ term. Of equal importance in defining an intrinsic state is the adjacency or topology of the atom in the molecule (Hall and Kier, 1990). Accordingly, the intrinsic state encodes two attributes:

1. The availability of the atom or group for intermolecular interaction (its potential for electronic interaction)
2. The manifold of bonds over which adjacent atoms may influence, and be influenced by, its state

The adjacency, encoded by the simple delta value δ, must therefore be a companion descriptor with the electronegativity in defining the intrinsic state. One possibility is to use the reciprocal of the adjacency, $1/\delta$, as an index of accessibility. The larger this value, the greater the accessibility of an atom or group. The product of the two terms produces an initial description of the intrinsic state value, I:

$$\frac{\delta^v - \delta}{\delta} \qquad (2.1)$$

The values for several hybrid atoms and groups are shown in Table 2.1. In this form, the intrinsic state, I, may be viewed as the ratio of the pi and lone-pair electron count to the count of avenues of intramolecular interaction, the number of sigma bonds in the skeleton for this atom. That is, due to the intramolecular interaction associated with the atom, pi and lone-pair electron density may be redistributed across the bonding network which is the set of sigma bonds in the molecular skeleton.

The value found for I in Eq. (2.1) for the carbon sp^3 atoms in each hydride group is zero since $\delta^v = \delta$. If we scale the $\delta^v - \delta$ term by one, the zero values are eliminated and there is a discrimination among the various hydrides of carbon arising from the different values of δ. This modification leads to Eq. (2.2):

$$\frac{\delta^v - \delta + 1}{\delta} \qquad (2.2)$$

This expression achieves the objective of encoding electronic structure, topology, and a fortuitous effect, an approximation of the valence state electronegativity. An alternative form of Eq. (2.2) reveals the pi and lone-pair count explicitly: $(\pi + n + 1)/(\sigma - h)$. A simplification of Eq. (2.2) can be made by adding 1 to the entire term in Eq. (2.2) to produce the intrinsic state for an atom or group in a molecule:

$$I = \frac{\delta^v + 1}{\delta} \qquad (2.3)$$

Table 2.3 shows the intrinsic states of second-row atoms and groups. There remains another consideration: the atoms and groups in a higher quantum level.

VII. HIGHER QUANTUM-LEVEL ATOMS

In considering covalently bonding atoms in higher rows of the periodic table, we note that electronegativity depends on the total electron count, Z, as well as

TABLE 2.3   Intrinsic State Values of Second-Row Hydrides

Atom hydride group	$I = [(\delta^v + 1)/\delta]$
>C<	1.250
>CH—	1.333
—CH$_2$—	1.500
>C=	1.667
—CH$_3$, =CH—, >N—	2.000
≡C—, —NH—	2.500
=CH$_2$, =N—	3.000
—O—	3.500
≡CH, —NH$_2$	4.000
=NH	5.000
≡N, —OH	6.000
=O	7.000
—F	8.000

the valence electron count, Z^v. The preceding derivation may be modified to deal with the attributes of higher level atoms and groups. This is evident since oxygen and sulfur have the same delta values by the current definition but they have different valence state electronegativities. The difference is even more pronounced when the halogen atoms are compared. Our choice of a modification to account for these differences is to focus on the δ^v in Eq. (2.3) since this carries the information about the valence electrons.

The principal quantum number, N, is inserted into Eq. (2.3) to reflect the influence of quantum level on the properties of valence electrons, leading to the difference in intrinsic states among different atoms. In order to retain the values derived from this equation for second-row atoms, some term including N should be used as a coefficient of the δ^v value. The value of $(2/N)\delta^v$ equals the value of δ^v when $N = 2$. Using an inverse square relationship to encode the influence of N on the diminished electronegativity of higher level atoms, the term $(2/N)^2$ was selected to modify the δ^v value in Eq. (2.3). The general equation for the intrinsic state of an atom or group becomes

$$I = \frac{(2/N)^2 \delta^v + 1}{\delta} \tag{2.4}$$

This modification results in an expression for the intrinsic state values for all atoms, including those in higher quantum levels shown in Table 2.4. The correlation of these I values with Mulliken–Jaffe electronegativity values has been reported by Kier and Hall (1981).

TABLE 2.4   Intrinsic States of Higher Quantum-Level Atoms

Atom	N	δ^v	δ	I
—SH	3	5	1	3.222
—S—	3	6	2	1.833
=S	3	6	1	3.667
—Cl	3	7	1	4.111
—Br	4	7	1	2.750
—I	5	7	1	2.120

VIII. FIELD INFLUENCES ON THE INTRINSIC STATE

We have derived the intrinsic state of an atom or group, but this expression does not reflect its position or influence within the field of other atoms in a molecule. The influence of the field and the surrounding molecular topology must be encoded into the atom description. This influence may take the form of a perturbation of the intrinsic state using some characteristic of every other atom in the molecule. A reasonable choice of attribute, producing a perturbation on an atom, is the intrinsic state of each other atom in the molecule. This approach is related to using electronegativities of other atoms to modify the state of each atom within the field of the overall molecular structure. The vehicle for this influence is the network of bonds linking each atom with all others in the molecule. This network is synonymous with the chemical graph model of the molecule over which electronegativity influence manifests itself.

A second consideration is the effect of separation of two atoms in a molecule on the influence each has on the intrinsic state of the other. Since the chemical graph is the model of the presence and connectivity of atoms within the molecule, the count of bonds or atoms in paths separating two atoms is one measure of the distance between them. This count was chosen for the unit of distance between any two atoms in a molecule. More precisely, the count of atoms in the minimum path length, r_{ij}, separating two atoms, i and j, is the distance selected to encode the influence between two atoms. Note that this count is equal to the usual graph distance, d_{ij}, plus 1: $r_{ij} = d_{ij} + 1$. From this model, we have chosen the difference between the intrinsic states of atom i and atom j ($I_i - I_j$) as the perturbation on each other. This effect is assumed to diminish by some power, m, of the distance; hence, the perturbation of I_i, called ΔI_{ij}, is expressed as

$$\Delta I_{ij} = \frac{I_i - I_j}{r_{ij}^m} \tag{2.5}$$

$$i\text{———}j(A)\text{—}j(B)\text{—}j(C)\text{—}j(D)$$

$$\Delta I_{ij}$$

r_{ij} m	2	3	4	5
1	1.00	0.67	0.50	0.40
2	0.50	0.22	0.13	0.08
3	0.25	0.07	0.03	0.02

FIGURE 2.2   Calculated ΔI_{ij} values for various r_{ij} and m values assuming $I_i - I_j = 2.000$.

The choice of the value of m results in a variable influence of distant and close atoms in the graph. Figure 2.2 shows the effect of m on the perturbation, ΔI_{ij}, and the effect due to the separation, r_{ij}. The value of m is a useful variable to modify the influence of distant or nearby atoms for particular studies. Most studies to date have employed a value of $m = 2$. To illustrate the calculation of a ΔI_{ij} value, Fig. 2.3 shows the steps in finding this value for a methyl group in ethyl acetate.

The total perturbation of atom i is a consequence of the influence of all other atoms in the molecule. Accordingly, the total perturbation of the intrinsic state of atom i should be a sum of these individual perturbations, $\Sigma_j \, \Delta I_{ij}$, a sum of all terms expressed in Eq. (2.6). The actual state of atom i in a molecule is the intrinsic state, I_i, plus the sum of all perturbations included in $\Sigma_j \, \Delta I_{ij}$. This bonded state of atom i is called the electrotopological state, S_i, and is expressed as

$$S_i = I_i + \Sigma_j \, \Delta I_{ij} \tag{2.6}$$

For brevity, the S_i term is called the E-State for atom i. The E-State index was introduced in a series of articles (Kier and Hall, 1990; Hall and Kier, 1991a,b; Kier et al., 1992).

IX. SAMPLE CALCULATIONS

To illustrate calculation of E-State values of a molecule, consider ethyl acetate using the hydrogen-suppressed graph and numbering the atoms (groups) in an arbitrary order. Beginning with the carbonyl oxygen atom, numbered 1, the

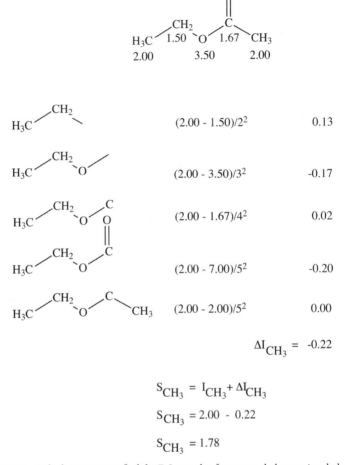

$$(2.00 - 1.50)/2^2 \qquad\qquad 0.13$$

$$(2.00 - 3.50)/3^2 \qquad\qquad -0.17$$

$$(2.00 - 1.67)/4^2 \qquad\qquad 0.02$$

$$(2.00 - 7.00)/5^2 \qquad\qquad -0.20$$

$$(2.00 - 2.00)/5^2 \qquad\qquad 0.00$$

$$\Delta I_{CH_3} = -0.22$$

$$S_{CH_3} = I_{CH_3} + \Delta I_{CH_3}$$

$$S_{CH_3} = 2.00 - 0.22$$

$$S_{CH_3} = 1.78$$

FIGURE 2.3 Calculation steps to find the E-State value for one methyl group in ethyl acetate.

several paths associated with this atom are identified by their intrinsic states (Fig. 2.4). In this case, there are five paths joining atom 1. Each path connected to atom 1 is coded for its influence on that atom using Eq. (2.5). For atom 1, the intrinsic state is 7.00. The ΔI_{ij} values associated with atom 1 are shown in Table 2.5. The sum of these ΔI_{ij} values is 2.82. This sum is the perturbation of the intrinsic state of atom 1, the carbonyl oxygen, by all other atoms. Using Eq. (2.6), the calculation of the E-State value for atom 1 is

$$S_1 = I_1 + \Sigma_j \, \Delta I_{1j} = 7.00 + 2.82 = 9.82$$

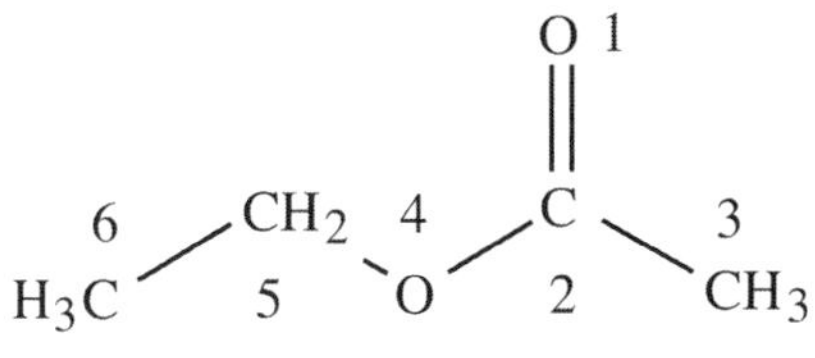

O 1
$\parallel$

6   CH$_2$ 4   C   3
H$_3$C   5   O   2   CH$_3$

Intrinsic State Values

$I(1) = 7.000$

$I(2) = 1.67$

$I(3) = 2.00$

$I(4) = 3.50$

$I(5) = 1.50$

$I(6) = 2.00$

FIGURE 2.4   Numbering scheme and intrinsic state values for ethyl acetate.

This process is repeated for each atom in ethyl acetate. The calculations can be presented as a ΔI_{ij} matrix for all atom pairs describing a path joining atom i. This matrix is shown in Table 2.6. The numbering scheme is the same as that in Figure 2.4. Note that the sum of ΔI_{ij} values for all atoms in the molecule is zero. All perturbations are confined to the molecule and are derived from the

TABLE 2.5   Calculated Fragment Values Leading from Atom 1 in Fig. 2.4

Remote atom no.	Remote atom I value	ΔI_{ij}
2	1.67	1.33
3	2.00	0.56
4	3.50	0.39
5	1.50	0.34
6	2.00	0.20
	$\Sigma \Delta I_{ij} = 2.82$	

TABLE 2.6 ΔI_{ij} Matrix for Atoms in Ethyl Acetate

Atom i	Atom j						$\Sigma_j \Delta I_{ij}$
	1	2	3	4	5	6	
1		1.33	0.56	0.39	0.34	0.20	2.82
2	-1.33		-0.08	-0.46	0.02	-0.02	-1.87
3	-0.56	0.08		-0.17	0.03	0.00	-0.62
4	-0.39	0.46	0.17		0.50	0.17	0.91
5	-0.34	-0.02	-0.03	-0.50		-0.13	-1.02
6	-0.20	0.02	0.00	-0.17	0.13		-0.22
							Sum = 0.00

atoms within the molecule. Some of the $\Sigma \Delta I_{ij}$ values are negative. In this case, the perturbation of that particular atom produces a loss in electron accessibility, reflected in a lower value of the E-State. Applying Eq. (2.6) to the data in Table 2.6, the values of the E-States for each atom in ethyl acetate can be calculated as shown in Fig. 2.5.

Several pieces of information are produced from these calculations. First, the sum of all E-State values in the molecule, ΣS_i, is equal to the sum of all intrinsic states, ΣI_i; that is, $\Sigma S_i = \Sigma I_i$. The calculation results in a reorganization of the electron accessibility as encoded in the intrinsic state values based on the relative electronegativity and topological environment of each atom. This corresponds to the reality that all valence electrons remain in a molecule despite perturbations imposed on them by internal reorganization. The atoms with greater electronegativities, such as the oxygen atom in this example, experience a net gain in their electron accessibility as indicated by an increase in their E-State values above their I values. In contrast, the sp^2 carbonyl carbon atom has an E-State value diminished from its intrinsic state because of its (i) lower electronegativity, (ii) proximity to higher electronegative atoms, and (iii) partially buried topological structure. These issues will be discussed in greater detail when the structure significance of the E-State is examined.

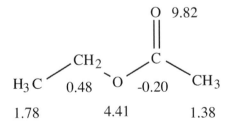

FIGURE 2.5   Calculated E-State values for ethyl acetate.

TABLE 2.7   Electrotopological State Calculations for N-Methylpropanamide Scheme (VIII)

Intrinsic state values

$I(1) = 2.000 \quad I(4) = 7.000$
$I(2) = 1.500 \quad I(5) = 2.500$
$I(3) = 1.667 \quad I(6) = 2.000$

$(I_i - I_j)/r_{ij}^2$ matrix

Atom i	Atom j						$\Sigma_j \Delta I_{ij}$
	1	2	3	4	5	6	
1		0.125	0.037	−0.312	−0.031	0.000	−0.181
2	−0.125		−0.042	−0.611	−0.111	−0.031	−0.920
3	−0.037	0.042		−1.333	−0.208	−0.037	−1.574
4	0.313	0.611	1.333		0.500	0.313	3.069
5	0.031	0.111	0.208	−0.500		0.125	−0.024
6	0.000	0.031	0.037	−0.313	−0.125		−0.369

A second observation concerns the two methyl groups. Although they are intrinsically equivalent if the molecule is dissected into its constituent atoms and groups, within the context of the complex system of the molecule, emergent attributes produce a difference between these structure features. This difference is encoded into the E-State values. The calculated difference is indeed significant, indicating that the role of the topology can be quite important in structure differentiation. The difference in E-State values is revealed in such properties as nuclear magnetic resonance. A second example of the calculation of the E-State using the N-methylpropanamide molecule, numbered as in Scheme VIII, gives the reader an opportunity to follow the steps described previously. This example is shown in Table 2.7.

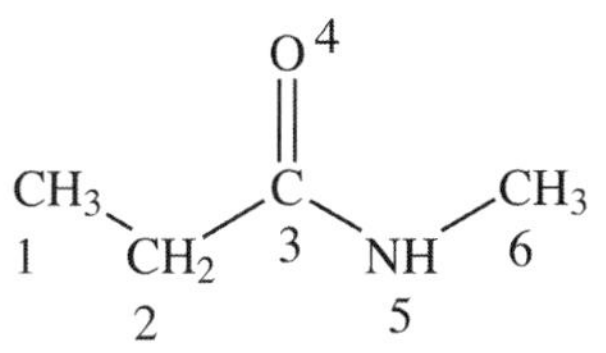

(VIII)

X. INFLUENCE OF STRUCTURE CHANGE ON E-STATE VALUES

The influence of structure on the E-State values may be revealed by examining several sets of calculations in which features are systematically varied. The influence of chain length on the terminal methyl group E-State is shown in Fig. 2.6. With lengthening of the chain the methyl group E-State value steadily increases, approaching a limiting value of 2.32. Figure 2.6 also shows that the methylene group adjacent to the terminal methyl group has a lower E-State value than any other group in the molecule. This methylene group has a limiting E-State value of about 1.38 in a very long chain molecule. The next methylene group in the molecule has a higher E-State value; each subsequent group has higher values until the most central methylene groups approach limiting values of 1.50. This value is equivalent to the methylene intrinsic state value.

The effect of branching in alkanes leads to significant perturbation of the intrinsic states of atoms near the branch points. In Fig. 2.7, several examples illustrate this effect. Methyl groups at a branch point have E-State values of approximately 2.22. A significant decrease in E-State value is found on the carbon at the branch point. Depending on the nature of the branching and the topology of the atoms involved, the carbon at the branch point may suffer a decline of 50% or more from its intrinsic state. The influence operating here is the availability of three or four adjacent atoms to draw from the intrinsic state as shown in Eq. (2.5). With three or four adjacent atoms of greater electronegativity, the branch point has multiple possibilities for the reduction of its E-State value. The effect of the topology is thus manifested in the creation of multiple ΔI_{ij} values operating on the branched atom.

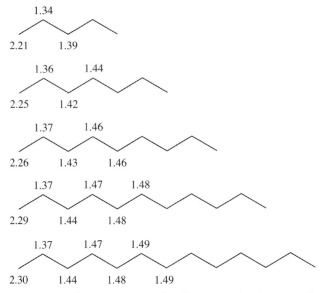

FIGURE 2.6   The changes in E-State values due to chain lengthening in alkanes.

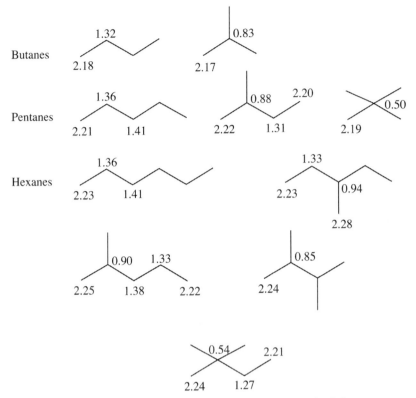

Butanes

Pentanes

Hexanes

FIGURE 2.7   Changes in E-State values due to branching in the skeleton.

The introduction of double and triple bonds in the alkane structure produces a significant perturbation in the E-State values of nearby atoms. The unsaturated atoms are also elevated in E-State value because of their higher intrinsic states. Some examples of these are shown in Fig. 2.8. Terminal unsaturation
results in higher E-State values for the sp² and sp atoms relative to the unsaturated atoms in midchain. Atoms near the double and triple bonds have E-States
lower than those in the corresponding saturated chains. An unsaturated atom
can produce significant changes in the E-State values of atoms up to three atoms away, where the E-State value may be reduced by 5–10%.

The introduction of a heteroatom into an alkane molecule produces an effect
on adjacent atoms commensurate with the I value of the heteroatom. Trends in
E-State values under the influence of heteroatoms are presented in Fig. 2.9. The
higher the I_i value of heteroatom i, the greater the E-State value of atom i and
the lower the E-State values of surrounding atoms. Closer atoms contribute

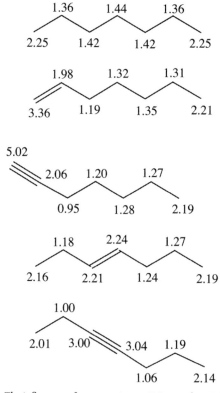

FIGURE 2.8   The influences of unsaturation on E-State values in hydrocarbons.

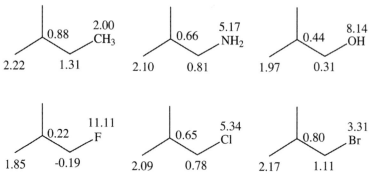

FIGURE 2.9   The influence of heteroatoms on the E-State values.

more to the heteroatom E-State and they lose correspondingly more from their intrinsic state values. The E-State values tend to reflect common chemical intuition regarding the structure of organic molecules.

XI. CALCULATION EXERCISES

The general algorithm for the calculation of E-State values has been introduced in this chapter, and examples have been provided to illustrate the calculation. In this section, the reader is presented with examples that illustrate several aspects of the calculation. It is intended that these calculations be carried out with the computer program, E-Calc, on the CD-ROM provided with this book. The style of this section is that of a workbook. Several simple problems will be presented to the reader for solving.

A. THE DELTA PARAMETER

The simple delta, δ, and the valence delta, δ^v, have been shown in this chapter to encode structure attributes used in defining the intrinsic states of an atom in a molecule. The reader is now asked to respond to several questions about these parameters.

1. The Simple Delta Value, δ

1. List the structure attributes encoded in the δ value.
2. Write down the δ values for each atom (or hydride group) in diethyl ether, toluene, and butyramide.
3. What is the numerical range possible for δ?
4. What is the topological significance of $1/\delta$?
5. What are the δ values for ethane, ethylene, and acetylene?

2. The Valence Delta Value, δ^v

1. Define the δ^v.
2. Define the δ^v for third-row atoms.
3. What structure information is encoded in δ^v?
4. Write down the δ^v for the atoms (hydride groups) in aniline, propanoic acid, and *ortho*-chlorophenol.
5. What information is common to both δ and δ^v for an atom?
6. What information is found in $\delta^v - \delta$ for an atom?

B. The Intrinsic State I

The intrinsic state value is derived from δ and δ^v values for an atom. There is a large amount of information in this parameter. A review in the form of questions is productive.

1. Derive I from δ and δ^v.
2. What two important pieces of information are encoded in I?
3. Calculate the I values for the atoms (hydride groups) in salicylic acid, alanine, and isobutane.
4. Calculate the I values for the halogen atoms.
5. What is the range possible for the I values?
6. What atoms or hydride groups have the same I value as that of a methyl group?

C. The E-State Value, S

The E-State value of an atom in a molecule is computed from the intrinsic state, the I value, and the difference between that I value and all other I values in the molecule. The differences, ΔI_{ij}, are computed as a function of the separations of every pair of atoms using the number of atoms in the path of separation as the distance. Two atoms bonded to each other are separated by one path denoted by two atoms. Two atoms separated by two bonds are separated by two paths denoted by three atoms. The sum of all ΔI_{ij} values impacting on an atom is the perturbation of that atom I value and is expressed as $I_i + \Sigma_j \Delta I_{ij}$. This expression is the E-State value, S, of the atom in consideration. A few questions focus on this derivation.

1. Compute the E-State values for the atoms in ethane, ethylene, and acetylene.
2. Compute the E-State values for the atoms in HCN, formaldehyde, and methylene chloride.
3. Compute the E-State value for the OH group in ethanolamine and for homologs of up to six methylene groups separating the OH from the NH_2 group. What is the shape of the curve relating $S(OH)$ and the number of C atoms in an homologous series?

D. The Direct Path Used
in the ΔI Calculations

The path between any two atoms used in the calculation of ΔI is chosen to be the shortest path possible. For example, in the heterocycle, oxazole (Scheme

TABLE 2.8 Number of Paths of Various Orders

	Order				
Molecule	1	2	3	4	5
Pentane	4	3	2	1	0
Isopentane	4	4	2	0	0
Neopentane	4	6	0	0	0
Aspirin	13	17	19	23	24
Quinine	27	39	56	76	105
Cholesterol	31	48	68	94	124
Morphine	25	40	62	95	138

IX), the path separating the oxygen atom and the nitrogen atom is by convention the two-bond path shown as a fragment in Scheme X. The alternative path shown in Scheme XI is not used to calculate ΔI. If two atoms are separated by two equal path lengths, then only one is used in the calculation of that ΔI value. For example, in the molecule, morpholine (Scheme XII), there are two paths connecting the O and the NH. Only one path is used to calculate the ΔI relating these two atoms. The following call attention to these calculation details:

1. Compute the E-State values for the two heteroatoms in morpholine.
2. Compute the E-State value for the meta position of the phenol molecule.

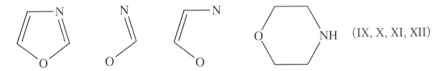

(IX, X, XI, XII)

E. Large Molecule Calculations

How large is a large molecule? A reasonable answer is one that produces errors with a modest rate of frequency when calculations such as the E-State values are made. It is easy to see how this can happen. The entries in Table 2.8 reveal the number of paths of various orders found for several molecules. It is reasonable to say that the first three molecules in Table 2.8 are simple enough to calculate by hand or with a pocket calculator. Try your luck calculating the E-State values for any of the other molecules. The obvious answer lies in a computer program that finds each path of each order and makes the computation rapidly. (The E-State calculations can be made with the program Molconn-Z.)[1]

[1]Molconn-Z may be obtained from Hall Associates Consulting, 2 Davis Street, Quincy, MA; SciVision, Inc., 200 Wheeler Rd., Burlington, MA 01803; Edusoft, LC, P.O. Box 1811, Ashland, VA 23005; and Tripos, Inc., 1699 South Hanley Road, St. Louis, MO 63144.

See the Appendix for instructions on use of the CD-ROM. The input for this program is accomplished through the use of a 2-D sketcher that allows easy drawing of the molecular structure.

EXERCISE 1

A sample calculation illustrates the steps in the process. The molecule chosen is N-methyl dopamine (Scheme XIII) using an arbitrary numbering scheme. Use the program as described in the Appendix. Examine the output, paying particular attention to the E-State values. From this output, answer the following questions:

1. Which atom has the highest E-State value?
2. Which atom has the lowest E-State value?
3. Which methylene group is most influenced by the molecular field?
4. Which ring carbon atom has the highest E-State value?
5. Repeat the calculation using a different numbering scheme.

(XIII)

EXERCISE 2

Do a similar exercise as that in Exercise 1 for the molecule N-methyl-propanamide. Answer questions 1, 2, 3, and 5 above.

EXERCISE 3

Calculate the E-State values for the two molecules formed by removal of the N-methyl group and by the addition of a second N-methyl group of the molecule in Exercise 1. Compare the E-State values of the nitrogen atoms in these two cases with the value from Exercise 2.

XII. SUMMARY

In this chapter, we have identified a molecule as a complex system in which the properties are more than the sum of atomic attributes. The atoms are not in isolation but constitute ingredients in a field in which each atom influences all others according to the intrinsic attributes of each. Field effects operate over the network of bonds that can be represented by a chemical graph. To encode the structure of each atom it is necessary to begin with a statement of the elementary attributes of the atom in a general case. We have identified this code as an intrinsic state which is a composite of the valence state electronegativity and the topology of the atom in a particular molecule. The intrinsic state of each atom is perturbed by the presence and connectedness of all other atoms in the molecule. An algorithm was developed to capture this perturbing influence. The result is an index of content and context of each atom in a molecule, called the E-State. Several examples reveal the influence of structure change on the numerical values of the indices.

XIII. A LOOK AHEAD

With the E-State derived and used in a few demonstration calculations, it is important to establish an idea of the significance and interpretation of the numerical values. In Chapter 3, we examine the information conveyed by the E-State indices for a variety of molecular types. A comparison will be made between the E-State and other nonempirical structure indices. An interesting relationship is developed with the free valence index from molecular orbital theory. The ability of the indices to differentiate among certain atomic properties will be demonstrated with nuclear magnetic resonance data. Finally, the spectrum of E-State values will be shown to describe the variation in the ability of molecular fragments to interact with other molecules.

REFERENCES

Dimoglo, A. S., Bersaker, I.B., and Gorbachov, M. Yu. (1988). Electron-topological study of structure–activity relationship of various inhibitors of α-chymotrypsin. *Khim-Farm. Zh.* **22**, 1355–1361.

Ghose. A. K., and Crippen, G. M. (1987). Atomic physicochemical parameters for 3-D structure directed QSAR 3. Modeling hydrophobic interactions. *J. Chem. Inf. Comput. Sci.* **27**, 21–26.

Hall, L. H., and Kier, L. B. (1977). A molecular connectivity study of electron density in alkanes. *Tetrahedron* **33**, 1953–1957.

Hall, L. H., and Kier, L. B. (1978). Molecular connectivity and substructure analysis. *J. Pharm. Sci.* **67**, 1743–1747.

Hall, L. H., and Kier, L. B. (1990). Determination of topological equivalence in molecular graphs from the topological state. *Quant. Structure–Activity Relat.* **9**, 115–131.

Hall, L. H., and Kier, L. B. (1991a). The electrotopological state: Structure information at the atomic level for molecular graphs. *J. Chem. Inf. Comput. Sci.* **31**, 76–83.

Hall, L. H., and Kier, L. B. (1991b). The electrotopological state: An atomic index for QSAR. *Quant. Structure–Activity Relat.* **10**, 43–48.

Hinze, J., and Jaffe, H. H. (1962). Electronegativity I. Orbital electronegativity of neutral atoms. *J. Am. Chem. Soc.* **84**, 540–549.

Kalizan, R., Osmialowski, K., and Halkiewicz, J. (1986). Quantum chemical parameters in correlation analysis of gas–liquid chromatographic retention indices of amines II. Topological electronic index. *J. Chromatogr.* **361**, 63–69.

Kier, L. B., and Hall, L. H. (1976). *Molecular Connectivity in Chemistry and Drug Research.* Academic Press, New York.

Kier, L. B., and Hall, L. H. (1981). Derivation and significance of valence molecular connectivity. *J. Pharm. Sci.* **70**, 583–587.

Kier, L. B., and Hall, L. H. (1986). *Molecular Connectivity in Structure–Activity Analysis.* Wiley, London.

Kier, L. B., and Hall, L. H. (1990). An electrotopological state index for atoms in molecules. *Pharm. Res.* **7**, 801–807.

Kier, L. B., Hall, L. H., and Frazer, J. W. (1992). An index of electrotopological state for atoms in molecules. *J. Math. Chem.* **7**, 229–237.

Klopman, G., and Raychandhury, C. (1988). A novel approach to the use of graph theory in structure–activity studies. Application of the qualitative evaluation of mutagenicity in a series of nonfused ring aromatic compounds. *J. Comp. Chem.* **9**, 232–243.

Mulliken, R. S. (1934). A new electroaffinity scale. *J. Chem. Phys.* **2**, 782–793.

Nilakantan, R., Bauman, N., Dixon, J. S., and Venkataraghavan, R. (1987). Topological torsion: A new molecular descriptor for SAR. *J. Chem. Inf. Comput. Sci.* **27**, 82–85.

Randic, M. (1984). On molecular identification numbers. *J. Chem. Inf. Comput. Sci.* **24**, 164–175.

Testa, B., and Kier, L. B. (1997). The concept of emergence–dissolvence in drug research. *Adv. Drug Res.* **30**, 1–14.

Testa, B., Kier, L. B., and Carrupt, P.-A. (1997). A systems approach to molecular structure, intermolecular recognition and emergence–dissolvence in medicinal research. *Med. Res. Rev.* **17**, 303–326.

CHAPTER 3

Significance and Interpretations of the E-State Indices

I. THE INTRINSIC STATE VALUES

The E-State indices have proven to be information rich as representations of molecular structure, especially in the exploration of relationships between activity and structure attributes (Joshi, 1993). In these biological encounters, intermolecular interactions play a fundamental role. Within the realm of intermolecular encounters and interactions, as are often found in ligand–effector engagements, the topological attribution may be of prime importance. Covalent bonding is a different phenomenon, in which major hybrid bonding changes, transitions among energy levels, and rearrangements are prominent. In the cases in which interaction and binding encounters prevail, the E-State indices, derived from the I values, foretell significant intermolecular interaction possibilities. Polar interactions are important in these cases. Usually, both electronic and topological factors come into play and should be considered together in the development of models of interactions. These types of information can be revealed in a series of analyses and comparisons presented in this chapter.

A. THE CARBON SP^3 HYDRIDES

An idea of the significance and interpretations of the intrinsic state I values may be gained by comparing subsets in Table 3.1. Consider the four hydrides of carbon sp^3 atoms, $-CH_3$, $-CH_2-$, $>CH-$, and $>C<$, which show a decreasing value of I. By definition, the I value encodes the electronegativity and the topological environment, as shown in the first form of Eq. (3.1):

$$I = \frac{(\delta^v - \delta) + 1}{\delta} + 1$$

$$= \frac{\pi + n + 1}{\delta} + 1 \qquad (3.1)$$

$$= \frac{\delta^v + 1}{\delta}$$

TABLE 3.1   Intrinsic State Values of
Second-Row Hydrides

Atom hydride group	$I = [(\delta^v + 1)/\delta]$
>C<	1.250
>CH—	1.333
—CH$_2$—	1.500
>C=	1.667
—CH$_3$, =CH—, >N—	2.000
≡C—, —NH—	2.500
=CH$_2$, =N—	3.000
—O—	3.500
≡CH, —NH$_2$	4.000
=NH	5.000
≡N, —OH	6.000
=O	7.000
—F	8.000

The electronegativities of Csp^3 and hydrogen are very similar among these four hydride states. Therefore, it can be concluded that the I values of the Csp^3 hydrides reflect a nearly pure influence of topology. The I values for this subset are 2.000, 1.500, 1.333, and 1.250, respectively. The lowest value, 1.250, for the quaternary carbon encodes a topologically buried atom which is virtually inaccessible to direct interaction with another molecule. This set of group values reveals how the E-State can encode the topological aspect of the structure influencing an interaction.

B. HETEROATOMS

The sp^3 hybrid terminal groups —F, —OH, —NH$_2$, and —CH$_3$ have decreasing I values of 8.00, 6.00, 4.00, and 2.00, respectively, correlating with their electronegativities. The relative ability of these groups to influence the E-State values of neighboring atoms is encoded in their intrinsic state values. Their potential for intermolecular interaction is also encoded in these I values, which in turn are manifested in the calculated E-State values. The same trend is observed among the sp^2 hybrid terminal groups =O, =NH, and =CH$_2$. The respective I values here are 7.00, 5.00, and 3.00. Again, the information encoded is the relative potential for intermolecular interaction as well as the influence on neighboring atoms. Each of these values is higher than for the corresponding carbon hydrides in the sp^3 hybrid set. This is due to the presence of an additional pi electron which corresponds to an increase in electronegativity.

Another instructive comparison is among the $=O$ and $—O—$ fragments, which have I values of 7.00 and 3.50, respectively. The lower value for the $—O—$ fragment is due to the topological state of the atom. Even so, molecular orbital calculations show comparable electron densities. The I values presage a lower potential for the $—O—$ fragment to engage in intermolecular interactions. This manifests itself in lower solubilities and boiling points of ethers relative to ketones of the same molecular weight. In fact, diethyl ether is soluble in water at about 6%, whereas methyl ethyl ketone is soluble at about 27%. The same influence of topology on properties is revealed in a comparison of the set of fragments $—NH_2$, $—NH—$, and $>N—$, which have I values of 4.00, 2.50, and 2.00, respectively. This numerical trend corresponds to a decrease in intermolecular interaction potential. A comparison of boiling points among molecules of equal molecular weight (comparable dispersion forces) with these three groups confirms this expectation as seen in the following series: hexylamine, 130°; dipropylamine, 110°; and triethylamine, 89°. The information in the topological state indices in this case is certainly quite significant in predicting the intermolecular interaction.

The accessibility of the atoms and groups discussed here may be interpreted as a probability of interaction with another molecule. In the case of atoms with high E-State values, this interaction could be acceptance of a hydrogen bond. The strength of the hydrogen bond relates closely to the E-State values in the series, $—F > =O > —OH > —N= > —NH_2$. The potential for donation of a hydrogen atom in a hydrogen bond is not directly encoded in the E-State. This development will be considered in Chapter 4 as an extension of the E-State formalism called the Hydrogen E-State.

II. COMPARISON OF THE E-STATE VALUES
WITH OTHER INDICES

The previous examples indicate that our derivation and calculation of the intrinsic state values provides a structure profile for an atom or group in a molecule that mirrors physical experience. The intrinsic state I values are derived as a unified attribution model containing information about the electronic and topological state of an atom or group in a molecule. The I values are not exclusively electronic or topological indices; they are a complex composite of these attributes. An idea of the nonequivalence of the intrinsic state (or E-State values) with other specialized indices may be developed by direct comparisons. In a series of studies by Joshi (1993), the relationships of the E-State values with charge and lipophilicity indices were revealed. Each of these comparisons is an effort to assess the E-State information relative to these isolated structure attributes.

TABLE 3.2   Gasteiger Charges and E-State Values
on Methyl Groups

R	R–CH₃	
	E-State	Gasteiger charge
CH_3	2.000	0.001
C_2H_5	2.125	0.003
n-C_3H_7	2.181	0.004
n-C_4H_9	2.212	0.004
i-C_4H_9	2.199	0.004
neo-C_5H_{11}	2.208	0.004
i-C_3H_7	2.167	0.006
i-C_4H_9	2.222	0.006
$(C_2H_5)_2CH$	2.278	0.007
$tert$-C_4H_9	2.188	0.009

A. ESTIMATED PARTIAL CHARGES

Gasteiger and Marsili (1978) created an iterative scheme to compute atomic charges based on partial equalization of orbital electronegativity using the Mulliken–Jaffe definition of orbital electronegativity (Mulliken, 1934; Hinze and Jaffe, 1962). Since the E-State index is based on the Kier–Hall electronegativity (Kier and Hall, 1981) in a scheme which is similar to electronegativity equalization, a comparison of E-State values to atomic charges calculated by the Gasteiger–Marsili method is informative. A group of linear and branched alkanes in Table 3.2 is used for comparing the atomic charges computed by the Gasteiger–Marsili method and the E-State values. Since the E-State values are defined for atom hydrides, the Gasteiger–Marsili charges for carbon atoms were added to charges for hydrogens attached to the carbon to obtain the total charge for the hydride. Correlation between the Gasteiger–Marsili group charge values and the E-State values was found not to be significant. This suggests that although both models use electronegativity-based perturbation of atomic properties as their basis, the E-State and the Gasteiger model characterization of molecular fragments are significantly different. In this case, an argument in favor of one model over another cannot be made because both models are based on different assumptions; however, both have been used successfully in the prediction of molecular properties.

An examination of the trend in the Gasteiger–Marsili atomic charges shows that the charge on the —CH_3 group in Table 3.2 increases along the sequence, methyl, ethyl, propyl, butyl, which is supported by chemical evidence as discussed by Fliszar (1983). This is a result of the use of Mulliken electronegativity values in the Gasteiger–Marsili scheme. A similar trend is observed for the E-State values. Since the Mulliken electronegativity of hydrogen is slightly less than that of Csp^3, replacing a hydrogen atom by a carbon atom results in a calculated negative inductive effect. The difference in the two methods is in the extensive explicit role of the topology in the E-State algorithm.

B. TAFT σ^* PARAMETERS

A comparison with Taft σ^* values (Taft, 1956) provides another look at atom-centered indices in an effort to further understand the E-State indices. The E-State values for the methyl group in molecules in Table 3.3 were compared with modified Taft σ^* values. The Taft σ^* value is an empirically derived constant which is defined as follows for a substituent group R:

$$\sigma^* = \frac{1}{2.48 \ [\log |k/k_0|_b - \log |k/k_0|_a]} \tag{3.2}$$

where k is the rate constant for the hydrolysis of an ester RCOOR and k_0 is the rate constant for the hydrolysis of the corresponding acetate ester CH_3COOR. The subscripts a and b refer to acid- and base-catalyzed reactions, respectively. The base-catalyzed hydrolysis is assumed to be influenced by both the steric and the electrostatic effects of the R group, whereas the acid-catalyzed hydrolysis is assumed to be sensitive predominantly to the steric effects. The difference between the two experimental values, given by the σ^* value, is thought to encode only the electronic influence of the R group. The Taft values are compared with the E-State values to check for any correspondence between them. The Taft σ^* values are also compared with the Gasteiger charges on the methyl group of alkane molecules listed in Table 3.2.

TABLE 3.3   Taft σ^* Values for the R Group

		R–CH$_3$	
No.	R	E-State	Taft σ^*
1	CH_3	2.000	0.000
2	C_2H_5	2.125	−0.100
3	$n\text{-}C_3H_7$	2.181	−0.115
4	$n\text{-}C_4H_9$	2.212	−0.124
5	$i\text{-}C_4H_9$	2.199	−0.129
6	$neo\text{-}C_5H_{11}$	2.208	−0.151
7	$i\text{-}C_3H_7$	2.167	−0.190
8	$sec\text{-}C_4H_9$	2.222	−0.210
9	$(C_2H_5)_2CH$	2.278	−0.225
10	$tert\text{-}C_4H_9$	2.188	−0.300

A comparison indicates only a moderate correlation between the Taft σ^* values and the E-State values. This lends additional support to the claim that the values from E-State characterization of molecular structure are different from the purely electronic descriptors. This argument is reinforced by the high correlation between the atomic charges calculated by the Gasteiger–Marsili method and the Taft σ^* values, both of which are offered as descriptors of the electronic characteristics of atoms in molecules.

C. REKKER F VALUES

In earlier work (Kier and Hall, 1993), it was suggested that the E-State values of groups may encode information about the type of noncovalent interactions an atom or atom hydride is most likely to encounter. Atoms and hydride groups may be arranged along a spectrum according to the corresponding E-State values found in actual molecules. Atoms and atom hydrides at the lower end of the spectrum are nonpolar and thought to engage primarily in dispersion types of interactions. Atoms at the lower end of the spectrum are buried and hence would be expected to have only limited dispersion interactions. Rekker (1977) proposed statistically derived weights for the contribution of atoms, their hydrides, and molecular fragments toward the partition coefficient of the molecule when partitioned between octanol and water. In general, atoms (or atom hydrides) with greater dispersion interactions have higher Rekker's f values. Since the E-State values characterize the polarity and the steric accessibility of an atom or atom hydride, the E-State values for atoms and atom hydrides (which do not engage in hydrogen bonding) are comparable to the Rekker f values.

Atoms and atom hydrides which are not commonly found in hydrogen-bonding interactions were chosen for this study. These atoms and hydrides are listed in Table 3.4. The f values for the sp^2 carbon hydrides were not reported in the literature and hence these hydrides were not included in the analysis.

Interactions between solute molecules and molecules in the organic phase are believed to be either hydrophobic interactions or dispersion interactions. In general, molecules with a greater proportion of nonpolar groups are more soluble in the organic phase and hence have a higher partition coefficient, P, usually expressed as the log P. By distributing the log P values of molecules among molecular fragments, the Rekker f values provide a measure of the dispersion interactions of individual molecular fragments. To an extent, these interactions are known to depend on both the accessibility of the solute atom to the solvent and the polarity and polarizability of an atom. The E-State values characterize both these features of atoms in molecules. Figure 3.1 shows the

TABLE 3.4   Rekker f Values for Atom and
Atom Hydrides

Atom type	Average E-State value	Rekker f value
>C<	1.249	0.200
>CH—	−0.287	0.235
>C=	0.345	[a]
—CH$_2$—	0.778	0.530
=CH—	1.786	[a]
—CH$_3$	1.860	0.702
—I	2.028	0.587
—Br	3.288	0.270
=CH$_2$	3.650	[a]
—Cl	5.863	0.061

[a] Rekker f value not available.

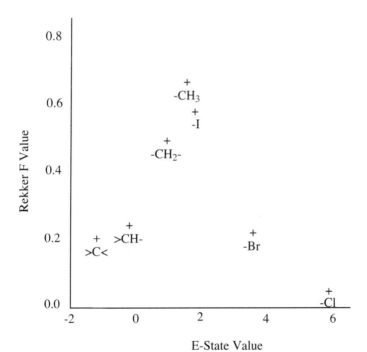

FIGURE 3.1   Plot of Rekker f values for a selected set of groups.

average E-State values of atoms which interact predominantly by dispersion-type interactions plotted versus the Rekker f values. The curve shows a peak at the position of the $-CH_3$ group. Atoms to the left of the peak show a progressive decrease in the f value (contributing to the log P of the molecule) because of restricted steric accessibility to the solvent molecules. Atoms to the right of the peak show a progressive decrease in the f value because of a higher polarity of the fragments. These results support the general interpretation of the E-State values as a unified approach to quantitate the steric and electrostatic structure of atoms in molecules.

D. COMPARISON OF LIPOPHILIC, STERIC, AND ELECTRONIC PARAMETERS

An extensive comparison of nonempirical structure parameters was reported by Waterbeemd et al. (1993). A set of 95 indices, including 6 E-State indices, were calculated for substituents on a pair of molecules selected to reflect an optimum diversity of structure. The first two principal components of each parameter are exhibited on a loadings plot as shown in Fig. 3.2. The parameters associated with the familiar attributes of lipohilic, steric, and electronic fall into three regions of the space. The E-State indices are found in a region of the space between and partially overlapping the electronic and the steric region. This finding is a further verification of the intent in the design of these indices—that they unify these attributes into the more fundamental state of atom–molecule structure. Clearly, the E-State indices are shown in Fig. 3.2 to be neither electronic nor steric parameters but rather a unification of these important structure attributes.

III. THE E-STATE SPECTRUM SIGNIFICANCE

The intrinsic state, I, and the resulting E-State values are not exclusively electronic or topological structure information; rather, they encode a unification of these attributes. Indeed, these attributes are manifestations of the same structure but are dissected by the chemist into convenient classes of phenomena that may be demonstrated separately. What do the intrinsic state and the E-State values encode? Some answers lie in a comparison of the numerical range of E-State values shown in Table 3.5. This table provides a ranking of the average E-State values for several atoms and fragments shown in Table 3.1. If we consider the relative type and extent of intermolecular interactions, we observe

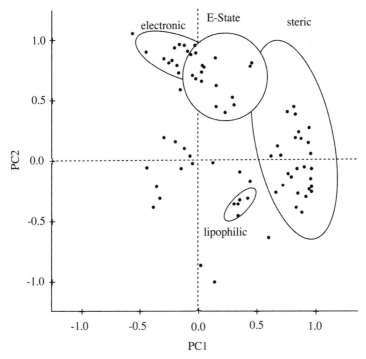

FIGURE 3.2   Plot of the first two principal components for several substituents for several different types of quantitative structure–activity relationship parameters.

that the stronger forces correspond to higher average E-State values. Strong electrostatic and hydrogen bond acceptance are attributes of $=O$, $-F$, $\equiv N$, and $-OH$ fragments. These typically have E-State values higher than 8.00 with a range up to the highest possible values. Below this set of atoms are a few groups which fall into a class in which bonding from forces classified as weak hydrogen bonding and dipolar forces occur. These include amine hydrides, halogens, carbon sp^2 hydrides, and ether oxygen. These fragments are not as readily available topologically and have a more modest electronegativity than the previous set. The range of E-State values for these fragments is approximately 3–8.

A set of fragments below the previous category comprises atoms that have a more modest electronegativity and may be bonded to more than one atom. This set includes $-CH_3$, $-CH_2-$, the higher weight halogens, and tertiary nitrogen. These are fragments that are commonly associated with van der Waals, dispersion, and the so-called hydrophobic bonding. The E-State range of these

TABLE 3.5   The E-State Spectrum and Interaction Forces

Average E-State value	Atom or Hydride	Interaction
14		H bonding
13		
12	—F	
11		Strong electrostatic
10	=O	
9	≡N, —OH	
8		
7	=NH	
6	—Cl	Dipolar, H bonding
5	—O—, =S, ≡CH, —NH$_2$	Weak electrostatic
4	=CH$_2$, =N—, —SH	
3	—Br, —NH—	
2	—CH$_3$, =CH—, I, ≡C—	van der Waals, dispersion,
1	—CH$_2$—, >N—, —S—	Lewis bases
0	>CH—, >C=	Buried
−1	>C<	Noninteractive

fragments is approximately 1.0–3.0. The lowest category on the average E-State scale is composed of fragments that have relatively low electronegativity and that are topologically buried, e.g., tertiary and quaternary carbons and fully substituted sp^2 carbon. The intermolecular interaction is very modest and highly proximity dependent. The quaternary carbon is a noninteractive atom. E-State values usually are below 1.0 for this category. This analysis indicates that the I values and the derived E-State values encode information about the potential of a fragment to engage in some form of intermolecular interaction.

A good test of the validity of the unification of attributes in the E-State is that of the correlation between the E-State index and the nuclear magnetic resonance (NMR) chemical shifts. This property reflects the environment of an atom in a molecule due to electronic and steric influences. Two studies have been reported (Kier and Hall, 1990; Hall et al., 1991) in which the E-State of oxygen atoms in a series of ethers and a series of carbonyl compounds was calculated. These indices were compared to the ^{17}O chemical shifts shown in Tables 3.6 and 3.7. The correlations between the E-State values and the chemical shifts (Δ) are very close.

For ethers:

$$\Delta = 92.56 \ S(—O—) - 441.65$$

$$r^2 = 0.99, \ s = 4.3, \ n = 10$$

TABLE 3.6   E-State of Ethers and ^{17}O NMR Chemical Shifts

Observed	Compound	$q(—O—)^a$	—O— E-State[b]	$^{17}O \Delta^c$	Calculated[d]
1	Dimethyl ether	−0.297	4.20	−52.2	−52.9
2	Ethyl methyl	−0.306	4.54	−22.5	−21.4
3	Isopropyl methyl	−0.313	4.75	−2.0	−2.0
4	t-Butyl methyl	−0.317	4.94	8.5	15.6
5	Diethyl	−0.315	4.83	6.5	5.4
6	Isopropyl ethyl	−0.322	5.04	28.0	24.9
7	t-Butyl ethyl	−0.325	5.23	40.5	42.4
8	Diisopropyl	−0.330	5.25	52.5	44.3
9	t-Butyl isopropyl	−0.333	5.44	62.5	61.9
10	Di-t-butyl	−0.334	5.63	76.0	79.5

[a] Oxygen partial charge computed with STO-3G basis.
[b] E-State values for ether oxygen.
[c] Measured ^{17}O NMR chemical shift (from C. Delseth and J.-P. Kintzinger, *Helv. Chim. Acta* **59**, 466, 1976).
[d] Calculated from regression equation: $\Delta = 92.56\ S(—O—) − 441.6$; $r^2 = 0.99, s = 4.3, n = 10$.

For carbonyls:

$$\Delta = -27.77\ S(=\!\!O) + 834.48$$

$$r^2 = 0.97,\ s = 3.67,\ n = 10$$

Clearly, the E-State indices encode relevant structure information influencing this property.

TABLE 3.7   E-State of Carbonyls and ^{17}O NMR Chemical Shifts

Observed	Compound	$q(=\!\!O)^a$	=O E-State[b]	$^{17}O \Delta^c$	Calculated[d]
1	CH_3CHO	−0.229	8.806	592.0	590.0
2	C_2H_5CHO	−0.230	9.174	579.5	579.7
3	$i\text{-}C_3H_7CHO$	−0.230	9.505	574.5	570.5
4	$(CH_3)_2CO$	−0.267	9.444	569.0	572.2
5	$CH_3COC_2H_5$	−0.270	9.813	557.5	562.0
6	$CH_3CO\text{-}i\text{-}C_3H_7$	−0.268	10.144	557.0	552.8
7	$(C_2H_5)_2CO$	−0.272	10.181	547.0	551.8
8	$C_2H_5CO\text{-}i\text{-}C_3H_5$	−0.271	10.512	543.5	542.6
9	$(i\text{-}C_3H_7)_2CO$	−0.274	10.843	535.0	533.4

[a] Oxygen partial charge computed with STO-3G basis.
[b] E-State values for ether oxygen.
[c] Measured ^{17}O NMR chemical shift (from C. Delseth and J.-P. Kintzinger, *Helv. Chim. Acta* **59**, 466, 1976).
[d] Calculated from regression equation: $\Delta = -27.77\ S(=\!\!O) + 834.48$; $r^2 = 0.99, s = 3.67, n = 9$.

IV. THE E-STATE AND FREE VALENCE

The consideration of the significance of the E-State index leads quite naturally to the concept of free valence introduced by Coulson as a molecular orbital description of potential reactivity at an atom (Coulson, 1947, 1948). This idea dates back to Thiele's partial valency and Werner's residual affinity. Free valence is defined as

$$F = N_{max} - N_r \tag{3.3}$$

where N_r is the sum of bond orders joining atom r. The constant, N_{max}, is the maximum possible value of N which is usually taken to be $3 + \sqrt{3}$ or 4.73 derived from the Huckel molecular orbital calculation of N_r as the central atom in trimethylenemethane (Scheme I).

$$
\begin{array}{c}
\bullet\, H_2C \diagdown \\
\diagup C{=\!=}CH_2 \\
\bullet\, H_2C
\end{array}
\tag{I}
$$

This N_{max} value arises from the bond order for the double bond calculated to be 1.73 plus an assumed value of 1.0 for each of the sigma bonds. An example of the results of this calculation is the molecule butadiene (Scheme II).

$$
\begin{array}{c}
H_2C{=\!=}CH{-\!-}CH{-\!-}CH_2 \\
0.89 \quad\; 0.45
\end{array}
\tag{II}
$$

The bond orders calculated using simple Huckel molecular orbital theory reveal a greater bond order for the 1,2 and 3,4 bonds. The total bond order values are estimated by assuming that all C–H and C–C sigma bonds have bond orders of 1.0. Summing the bond orders at each atom then gives 3.89 for the 1 and 4 position atoms and 4.34 for the 2 and 3 position atoms. The free valence values are then calculated from Eq. (3.3) to be 0.89 for the 1 and 4 positions and 0.45 for the 2 and 3 positions.

It can be interpreted that this free valence encodes the relative amount of leftover or residual bonding capability available for intermolecular interactions which is the potential reactivity at a particular atom. In the case of butadiene, the free valence correctly predicts the atom that is the most reactive to neutral reagents. Many examples of the use of free valence to predict reactivity in aromatic hydrocarbons have been reviewed (Streitwieser, 1961; Gutman, 1978). It is generally believed that free valency is related to the forces initiating a reaction—forces which are largely due to residual affinity for approaching atoms rather than to strong interactions leading to transition state perturbations. These affinities permit the formation of incipient bonds without any severe deformation of the reagents. It is clear that in modern terms, we are looking at an index which mirrors the phenomena associated with receptor or protein noncovalent bonding.

As more sophisticated all-valence electron M.O. methods emerged in the 1960s, the concept of free valence was largely shelved in favor of population analyses, frontier orbitals, and electrostatic potentials to predict the relative reactivity of parts of molecules. The shortcoming of the classical free valence formulation is that it was derived for conjugated hydrocarbon systems; its extension to atoms with lone-pair electrons was never made. If we can find a relationship between E-State and free valence of conjugated molecules, we may conclude that the E-State is comparable to an extended free valence index but its extension includes sp^3 hybrid carbon atoms and heteroatoms. Accordingly, we have examined and reported on several polycyclic aromatic hydrocarbons with computed free valences in an effort to find possible correlations (Kier and Hall, 1997).

A. COMPARISON OF FREE VALENCE
AND E-STATE VALUES

To determine if there is a significant relationship between the free valence and the corresponding E-State values, we have extracted free valence values from a compilation of Huckel molecular orbital calculations by Coulson and Streitwieser (1962) of a wide variety of conjugated carbon atoms shown in Table 3.8. A comparison of these values with the computed E-State values is shown. It is evident from the correlation, $r = 0.92$, that there is a significant relationship among these free valencies and the E-State values. Figure 3.3 shows a plot of free valence values versus E-State values. The symbols shown in the plot indicate the six different types of unsaturated carbon atoms. The letter C indicates the location of each symbol and corresponds to the average value for the several examples of each type found in the data set.

B. FREE VALENCE INDICES
FOR UNSPECIFIED ATOMS

The free valence has been computed for a series of unspecified atoms attached to a general conjugated system $C = C - X$, where X may be a heteroatom contributing pi and/or lone-pair orbital electrons to the system. The inclusion

TABLE 3.8 Free Valence and E-State Values for Hydrocarbons

Atom No.	Molecular Structure	Atom ID	Group	Free valence	E-State value
1		a	$CH_2=$	0.84	3.36
2		b	$=CH-$	0.39	1.64
3		a	$CH_2=$	0.86	3.46
4		b	$=CH-$	0.38	1.71
5		c	$=CH-$	0.46	1.83
6		a	$CH_2=$	0.91	3.56
7		b	$=C<$	0.10	0.87
8		c	$=CH-$	0.42	1.66
9		d	$CH_2=$	0.82	3.46
10		a	$CH_2=$	0.88	3.52
11		b	$=CH-$	0.37	1.71
12		c	$=CH-$	0.52	1.85
13		d	$=C<$	0.15	1.00
14		e	$=CH-$	0.41	1.73
15		f	$CH_2=$	0.84	3.56
16		a	$CH_2=$	0.83	3.52
17		b	$=CH-$	0.41	1.66
18		c	$=C<$	0.12	0.83
19		d	$CH_2=$	0.91	3.66
20		a	$CH_2=$	0.85	3.68
21		b	$=CH-$	0.40	1.85
22		c	$=C<$	0.19	0.96
23		a	aCHa	0.44	2.02
24		b	aCHa	0.39	2.00
25		c	aCHa	0.41	1.99
26		d	aCas	0.11	1.17
27		e	$=CH-$	0.41	1.83
28		f	$CH_2=$	0.82	3.63
29		a	aCHa	0.45	2.10
30		b	aCHa	0.39	2.05
31		c	aCHa	0.42	2.03
32		d	aCas	0.10	1.23
33		e	$=CH-$	0.48	2.12
34		a	aCHa	0.44	2.08
35		b	aCHa	0.40	2.03
36		c	aCHa	0.41	2.02
37		d	aCas	0.12	1.18
38		e	$=C<$	0.14	1.08
39		f	$CH_2=$	0.89	4.10

TABLE 3.8 *(continued)*

Atom No.	Molecular Structure	Atom ID	Group	Free valence	E-State value
40	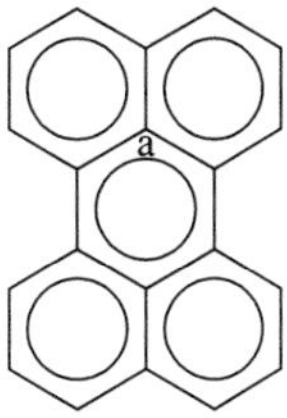	a	aaCa	0.15	1.39
41	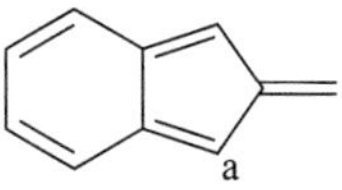	a	=CH—	0.59	2.09
42	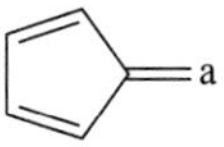	a	CH$_2$=	0.97	3.68
43	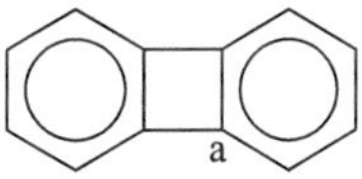	a	aaCa	0.22	1.40
44	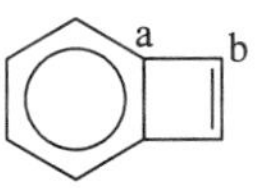	a	aaCa	0.24	1.36
45		b	=C<	0.62	2.12
46	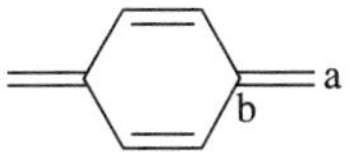	a	CH$_2$=	0.97	3.73
47		b	=C<	0.07	1.04
48	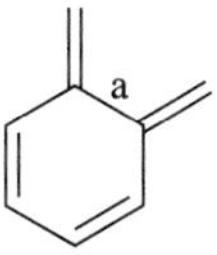	a	=C<	0.09	1.01
49	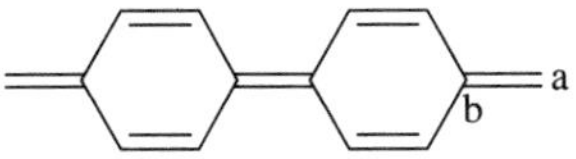	a	CH$_2$=	1.02	3.86
50		b	=C<	0.06	1.04
51	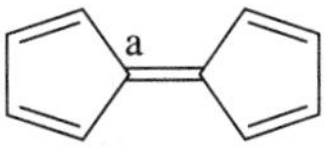	a	=C<	0.24	1.31
52	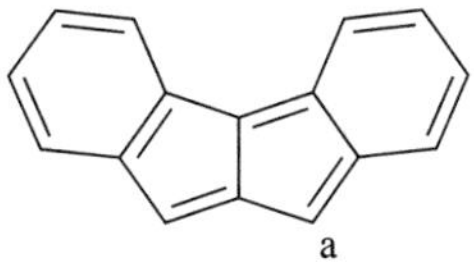	a	=CH—	0.62	2.28

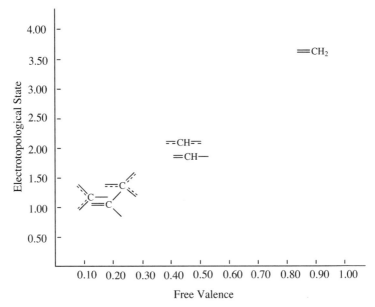

FIGURE 3.3 Plot of E-State values versus free valence values for the six types of unsaturated carbon atoms.

of atoms or groups such as these has been treated by assuming that their influence on the conjugated system is due to their variable electronegativity. This has been encoded into the Huckel M.O. calculation as parameters modifying the Coulomb integral, α, and the resonance integral, β (Kier, 1971; Coulson and Streitwieser, 1962). In Huckel M.O. calculations, the Coulomb integral is an approximation of the attraction between a pi electron and the atom core. It is expressed in this case by employing a parameter, α, encoding the relative electronegativity of the unspecified atom X. The parameters are h, modifying the Coulomb integral, and k, modifying the resonance integral. The modifiers take the form:

$$\alpha_x = \alpha_C + h\beta \quad \text{and} \quad \beta_{CX} = k\beta$$

The calculated free valence values for a variety of X groups are shown in Table 3.9. As the electronegativity increases with an increase in h, the free valence index also increases. This clearly establishes a relationship between this electronic attribute for an atom and the free valence index. In a study comparing E-State values and free valence, we found a strong relationship. From this we surmised that at a level of structure in which free valence can be calculated,

TABLE 3.9   Free Valence and Huckel M.O. Coulomb Integral

$$C = C - X$$

h_x	k_{cx}	F_r
0.00	0.50	1.28
0.50	0.50	1.41
1.00	0.50	1.48
0.00	1.00	1.02
0.50	1.00	1.15
1.00	1.00	1.25
0.00	1.50	0.90
0.50	1.50	0.98
1.00	1.50	1.07

there is a strong parallel with E-State values. The free valence calculations were introduced to account for pi electron influences in conjugated systems. At this level, the two indices portray a common picture when the numerical values are compared for the same molecules.

From a theoretical consideration, the perturbed intrinsic state

$$\Sigma_j \, \Delta I_{ij} = \Sigma_j \, \frac{I_i - I_j}{r_{ij}^2} \qquad (3.4)$$

and the summed bond orders impinging on an atom in Eq. (3.3) convey similar information. Both are derived from relative electronegativities of bonded atoms, manifesting a polarity between the two. The cumulative effect of these polarity differences results in a certain amount of nonbonded electron content at each atom. In the E-State formalism, all valence electrons are included in the electro-negativity calculation; hence, the derived intrinsic state value contains the non-bonded part of the electron component. In the case of the free valence, the calculation is built around the pi electrons resident in a conjugated system. The other valence electrons, the lone pairs, are only implicit in the modification of the carbon Coulomb integral used to characterize a heteroatom. In the free valence calculations, the sigma bonds are counted as contributing a value of one in the summation at each atom.

It can now be stated that all the valence electrons in the E-State calculation are integrated into the E-State formalism in such a way that separate terms are not necessary for the estimation of their contributions. As a result, the relative availability of the electrons accessible to intermolecular interactions is quantitated more explicitly with the E-State index than has been possible to achieve with the free valence as originally formulated.

V. SUMMARY

The E-State is derived as a composite index embracing both electronic and steric attributes of atoms in molecules. This combined information is shown in this chapter to be resident in the indices when selected atoms and groups are compared. The point is made that the E-State indices are not directly comparable to molecular orbital-type indices or to steric parameters. The E-State is demonstrated by several comparisons to be a unified attribution of an atom, reflecting its electronegativity, the electronegativity of proximal and distal atoms, and its topological state. The intriguing notion that the E-State may be similar to the free valence has been outlined based on recent work (Kier and Hall, 1997).

VI. A LOOK AHEAD

The basic structure of the E-State paradigm has been derived, exhibited, and related to existing atom and molecular parameters. In Chapter 4, some extensions of the E-State method will be developed and evaluated. These build on the understanding of the E-State derived from Chapters 2 and 3.

REFERENCES

Coulson, C. A. (1947). The theory of the structure of free radicals. *Discussions Faraday Soc.* 2, 9–18.

Coulson, C. A. (1948). *Valence*. Oxford Univ. Press, Oxford.

Coulson, C. A., and Streitwieser, A. (1962). *Dictionary of Pi-Electron Calculations*. Freeman, San Francisco.

Fliszar, S. (1983). *Charge Distribution and Chemical Effects*. Springer-Verlag, New York.

Gasteiger, J., and Marsili, M. (1978). A new model for calculating atomic charges in molecules. *Tet. Lett.* 34, 3181–3184.

Gutman, I. (1978). Topological formulas for free-valence index. *Croatica Chem. Acta* 51, 29–33.

Hall, L. H., Mohney, B. K., and Kier, L. B. (1991). The electrotopological state: An atom index for QSAR. *Quant. Structure–Activity Relat.* 10, 43–51.

Hinze, J., and Jaffe, H. H. (1962). Electronegativity I. Orbital electronegativity of neutral atoms. *J. Am. Chem. Soc.* 84, 540–549.

Joshi, N. (1993). Atom focused parameters for the electronic and topological attributes of atoms in molecules. Doctoral dissertation, Virginia Commonwealth University, Richmond.

Kier, L. B. (1971). *Molecular Orbital Theory in Drug Research*. Academic Press, New York.

Kier, L. B., and Hall, L. H. (1981). Derivation and significance of valence molecular connectivity. *J. Pharm. Sci.* 70, 583–587.

Kier, L. B., and Hall, L. H. (1990). An electrotopological state index for atoms in molecules. *Pharm. Res.* 7, 801–807.

Kier, L. B., and Hall, L. H. (1993). An atom centered index for drug QSAR models. *Adv. Drug Res.* **22**, 1–38.

Kier, L. B., and Hall, L. H. (1997). Free valence and the E-State values. *J. Chem. Inf. Comput. Sci.* **37**, 548–552.

Mulliken, R. S. (1934). A new electroaffinity scale. *J. Chem. Phys.* **2**, 782–793.

Rekker, R. (1977). *The Hydrophobic Fragment Constant.* Elsevier, Amsterdam.

Streitwieser, A. (1961). *Molecular Orbital Theory for Organic Chemists,* pp. 329–341. Wiley, New York.

Taft, R. W. (1956). Separation of polar, steric, and resonance effects in reactivity. In *Steric Effects in Organic Chemistry* (M. Newman, Ed.), pp. 556–675. Wiley, New York.

Waterbeemd, H., Carrupt, P.-A., Testa, B., and Kier, L. B. (1993). Multivariate data modeling of new steric, topological and CoMFA-derived substituent parameters. In *Trends in QSAR and Molecular Modeling* (C. G. Wermuth, Ed.), pp. 69–75. Escom, Leiden, The Netherlands.

CHAPTER 4

Extended Forms of the E-State

The E-State formalism, as derived and demonstrated previously, represents the individual atoms (or hydride groups) in a molecule. It is appropriate to consider modifications and additions to the basic model. In this chapter, we explore several of these extensions. Some of these are in the early stages of development but worthy of consideration; others are further along and have been used in studies. All of these topics lay the groundwork for further study and application.

I. SPECIFIC CONSIDERATION OF HYDROGEN

Since the E-State values are derived from a hydrogen-suppressed graph, the numerical values of groups subsume the contributions from the separate atoms into a single index. This may be a tolerable situation in the case of the alkanes in which the carbon and hydrogen atoms are very close in electronegativity. There is little appreciable polarity in the C–H bonds and so the group concept is a legitimate consolidation of the two atoms. On the other hand, representation of these nonpolar (lipophilic) hydrogen atoms may require a separate and independent encoding for modeling purposes. It is clearly the case that highly polar groups should be treated separately, such as the hydroxyl group, —OH, in which the difference in electronegativity between the two atoms is high. The H atom in this case, which is part of a polar bond, is much more reactive and interactive than in the case of an alkane.

The H atom in some applications of the E-State requires an independent consideration of its electrotopological state. This independence is demonstrated in situations in which a hydrogen bond may form or an intermolecular interaction involving separate atoms in a group is to be considered in a structure–activity model. In these situations, the strength of these interactions is dependent on the structure of the individual atoms in the group. Each atom, however, plays a different role and so the standard E-State group treatment described earlier falls short of a useful E-State description of the hydrogen atoms. Improvements have been explored and are presented here. We now consider three approaches to this issue.

A. THE HYDROGEN-EQUIVALENT MODEL

In this model, the hydrogen atom of the hydroxyl group is considered to be an atom with its own electronic interaction and topological state, as is the case for any other atom in the molecule. For example, a molecule such as ethanol is represented in Scheme I.

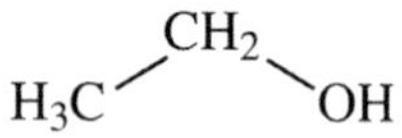

$$H_3C \diagup \overset{CH_2}{} \diagdown OH \qquad (I)$$

Using the algorithm for the intrinsic state, I, the calculated intrinsic state of the hydroxyl hydrogen is derived from

$$I(H) = \frac{(2/N)^2 \delta^v + 1}{\delta} \qquad (4.1)$$

where N is the principal quantum number, Z is the valence electron count, δ is the adjacency or sigma electron count, H is the attached hydrogens, and $\delta^v = Z - H$. Since $Z = 1$, $H = 0$, $N = 1$, and $\delta = 1$ for hydrogen, it follows that $I(H) = 5$. For the other atoms in ethanol, $I(CH_3) = 2$, $I(CH_2) = 1.5$, and $I(-O-) = 3.5$. The E-States calculated for the atoms in ethanol using these values are shown in Scheme II along with the H value for O–H from Eq. (4.1).

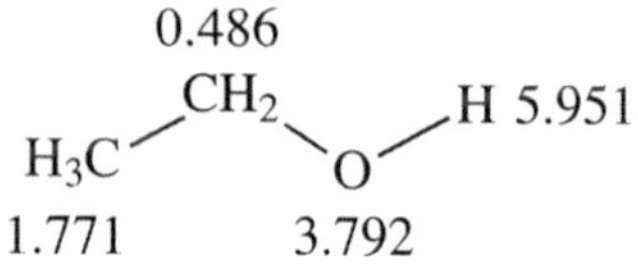

$$\overset{0.486}{\underset{1.771}{H_3C}} \diagup \overset{CH_2}{} \diagdown \underset{3.792}{O} \diagup \overset{H\ 5.951}{} \qquad (II)$$

For acetic acid we calculate the values shown in Scheme III. From a preliminary inspection we can say that the H in the OH group of the alcohol is more electrotopologically accessible. We may infer that this functional group is more available for hydrogen bonding. If we conclude that the difference in E-State values between two bonded atoms reflects the polarity of that bond, then a comparison of the two O–H bonds is revealing for ethanol:

$$S(H) - S(O) = 2.159 \qquad (4.2)$$

and for acetic acid:

$$S(H) - S(O) = 2.350 \qquad (4.3)$$

$$\begin{array}{c} \text{O 9.403} \\ \| \\ \underset{\underset{1.174}{H_3C}\ -0.578}{C} \diagdown \underset{3.361}{O} \diagup {}^{H\ 5.711} \end{array} \qquad \text{(III)}$$

The model indicates that the acid has a more polar O–H bond, in agreement with experimental evidence. A problem arises, however, if we extend this model to ethylamine. In this case, the $S(H)$ is 6.326 for each hydrogen atom and the $S(N)$ is 0.625. Clearly, the difference in these E-State values, 5.701 (twice that of O–H), does not accurately reflect the relative polarities. The E-State values of the hydrogens in ethylamine suggest much stronger hydrogen bonding potential than that for the alcohol. This result is not in agreement with experimental evidence. It appears that this line of reasoning is not fruitful for parameterizing the hydrogen atoms in specific cases.

B. THE PARTITION MODEL

In this model, the $I(H)$ is derived as a function of the atom to which it is attached. One approach is to calculate the $S(OH)$ value as in current practice and then to partition it between the O and the H atoms. For ethanol, the E-State values calculated are shown in Scheme IV. The $S(OH)$ of 7.569 is now partitioned among the O and the H atoms in a way that reflects experience and is also in harmony with the E-State method. Of course, the calculations for alkanes should yield $S(H)$ values that reflect their chemical and interaction potential. No scheme for partitioning has been proposed.

$$\begin{array}{c} 0.250 \\ CH_2 \\ \underset{1.681}{H_3C} \diagup \quad \diagdown \underset{7.569}{OH} \end{array} \qquad \text{(IV)}$$

C. A HYDROGEN INTRINSIC STATE
USING AN X–H MODEL

In this method, the focus is directly on the hydrogen atom where only the influence from the attached atom is considered. Since X and H are the only atoms considered, the influence on H must arise from some attribute of X, the electronegativity with a small input from topology. From the original expression

TABLE 4.1   Hydrogen Intrinsic States from the
X–H Model

X–H	I(H)	$\delta^v - \delta$	$(\delta^v - \delta)^2/\delta$
—OH	6.00	4	16
=NH	5.00	3	9
—NH$_2$	4.00	2	4
≡CH	4.00	2	4
=CH$_2$	3.00	1	1
—NH—	2.50	2	2
=CH—	2.00	1	1
—CH$_3$	2.00	0	0
—CH$_2$—	1.50	0	0
—CH<	1.33	0	0

quantifying the Kier–Hall electronegativity, the $(\delta^v - \delta)$ term is used as a start-ing point. For this application, the electronegativity influence is assumed to be accentuated and so we adopt the expression $(\delta^v - \delta)^2/\delta$ as a descriptor for X in X–H fragments. Table 4.1 shows the trends in the numerical results.

These values rank and quantify the H-donor ability of various X–H groups. It is intuitively reasonable and agrees with some molecular orbital calculations of water hydrogen bonding to various groups (Jhon and Sheraga, 1988; Wee et al., 1990). The OH group is favored to donate a hydrogen; an imine is next with NH$_2$ and a ≡CH group is a distant third. Note that the values for alkane CHs are zero. These values parallel E-State index values, which are a measure of the electron richness and accessibility but not a measure of hydrogen lability. This new value is exclusively for hydrogen atoms. When blended with the stan-dard S values, there results a ranking of both donor and acceptor values for functional groups, as shown in Table 4.2. These new $S(H)$ values, separately

TABLE 4.2   Blending of Donor–Acceptor
H-Bond Values

X–H	S(X) or S(H)
—OH (donor)	16
—F (acceptor)	~14
=O (acceptor)	~12
—OH (acceptor)	~10
=NH (donor)	9
—NH$_2$ (acceptor)	~4
—NH$_2$ (donor)	4

computed, may be used with conventional E-State values to give a composite picture of the donor–acceptor properties of functional groups. See Section II for an application.

D. ELECTRONEGATIVITY-BASED HYDROGEN E-STATE

In an another approach to quantifying hydrogen atoms, Hall and Vaughn (1997) focused on the electronegativity of each atom or hydride group. All hydrogen atoms encountered in organic molecules occupy a terminal position; that is, a topology described with a δ value of one: $\delta = 1$. Topology is not a crucial factor in determining the E-State value for a hydrogen atom. No matter whether the hydrogen atom is part of a hydride group in which the X–H bond is polar or nonpolar, relative electronegativity plays the dominant role in determining the character of this bond. A new formalism can be based on the same formalism as introduced in Chapter 2 with one exception. The intrinsic state I value for the hydrogen atom is based entirely on the relative electronegativity of the attached atom, X (since $\delta = 1$). The Kier–Hall relative electronegativity (KHE) is used to describe the electronegativity of each atom or hydride group in the molecule (Kier and Hall, 1981). KHE is defined as follows:

$$KHE = \frac{\pi + n}{N^2} \tag{4.4}$$

where π is the number of electrons assigned to pi orbitals, n is the number of electrons assigned to lone-pair orbitals, and N is the principal quantum number of the atom. For hydrogen atoms, KHE is taken to be -0.200. (In the current version of the Molconn-Z program, this algorithm is used for computation of hydrogen E-State values.)

This formalism for the hydrogen E-State has been applied in quantitative structure–activity relationship (QSAR) studies. Hall and Vaughn (1997) analyzed the toxicity of phenols to *Tetrahymena pyriformis*. A four-variable model was developed and is described in detail in Chapter 8. One of the variables is the E-State index for the phenolic hydrogen atom, which accounts for ~20% of the calculated toxicity of the 50 phenols in the data set. Small numerical values for the —OH hydrogen E-State index indicate low electron density and hence high polarity. Analysis of the QSAR equation shows that the greater the polarity of the phenolic OH, the greater the toxicity. In a second investigation, Gough and Hall (1998) examined the toxicity of amide herbicides on rats. A five-variable model was found to give a good account of the toxicity of the 50 compounds in the data set. Furthermore, predictions on 9 compounds not in the data set were very good. Two hydrogen E-State indices were found in the QSAR equation, including the hydrogen E-State for —OH groups.

A second hydrogen E-State index was calculated for all nonpolar hydrogen atoms and given the symbol SHother. This variable is an example of an atom-type E-State index which is described later. It represents the contribution of hydride groups which may participate in hydrophobic-like interactions. A detailed analysis of this study is given in Chapter 8.

II. THE E-STATE AS A THREE-DIMENSIONAL FIELD

A. THREE-DIMENSIONAL QSAR

Three-dimensional (3-D) QSAR studies are a relatively new branch of chemical information sciences that combine statistical analysis of molecular data with sets of 3-D molecular property descriptors. The prototype of this class of technique is comparative molecular field analysis (CoMFA) as described by Cramer *et al.* (1988). In CoMFA, molecules are described by fields, which are in practice numerical maps on evenly spaced grid points. The map value at any grid point is the calculated effect of all the atoms in the molecule over the distance to the grid point which can be envisioned as a "test atom" or "device." This calculation requires that each atom in the molecule have some atomistic property and that there is a known range response of this property. For example, an electrostatic field is represented by an atomic charge as the property and Coulomb's law, $E = q_1 q_2 / r_{12}^2$, as the function of distance, where q_1 is the atomic charge, q_2 is the charge on the test atom (grid point), and r_{12} is the distance between the two points. In fact, the original CoMFA procedure used this electrostatic model and a steric field representing shape that was calculated using the Lennard–Jones potential function. To obtain models, the values of the field points are used as the independent variables in partial least squares (PLS) regressions that are cross-validated. Impressive results, in terms of creation of predictive models for biological activity, have been obtained in many research laboratories using CoMFA with these two fields (Welch *et al.*, 1994; Kroemer *et al.*, 1995; Hocart *et al.*, 1995; Glennon *et al.*, 1994).

Other properties are potentially useful as 3-D fields, and several have been reported: hydrophobic, hydrogen bond, molecular orbital, etc. (Waller and Kellogg, 1996). The E-State and Hydrogen E-State (HE-State) are particularly attractive for the creation of a field because they are nonempirical and because they encode simply the electronegativity of the atom, the inductive influence of other atoms in the chemical graph, together with its topological state. These are attributes that are otherwise difficult to assess. While the distance response function of fields with known physical descriptions (e.g., electrostatic) are based on the associated interaction laws of the property, the responses for less well-understood properties such as hydrophobicity or nonempirical structure attributes such as the E-State are in the early stages of development and are currently handled pragmatically.

B. CALCULATION OF E-STATE FIELDS

Kellogg *et al.* (1996) have shown how to calculate 3-D E-State fields. The value, E, at each grid point (t) is represented as

$$E_t = S_i \, f(r_{ij})$$

over all atoms, i, in the molecule, where S_i is the E-State (or HE-State) for atom i, and $f(r_{ij})$ is a function of the distance between the atom and the grid point t. A variety of functional forms for the distance behavior, $f(r)$, can be explored, including $1/r$, $1/r^2$, $1/r^3$, $1/r^4$, and e^{-r}. Since the $1/r^n$ functions are discontinuous at grid points close to atoms, field default values (e.g., zero) should be set for grid points within one van der Waals radius of any atom in the molecule.

C. DISPLAY OF 3-D E-STATE FIELDS

The E-State field maps can be contoured and displayed in a manner analogous to the molecular electrostatic potential and other properties. The most significant difference between the E-State and HE-State fields is that the E-State is localized on heavy (nonhydrogen) atoms, whereas the HE-State is localized on the hydrogens. Thus, these two fields are particularly complementary and the HE-State field will probably not normally be highly correlated with the E-State field.

D. ADVANTAGES OF FIELD REPRESENTATION OF E-STATE IN QSAR

While the E-State is an atomistic parameter with significant applications for conventional QSAR, its utility has some limitations. First, the effects of chemical substitutions and structure differences must be inferred from the examination of parameters on common atoms rather than on the atoms of the substituents. This, by necessity, requires fairly homogeneous sets of molecules in the

QSAR analyses. This limitation is addressed and circumvented by using the atom-type E-State indices described later and illustrated in Chapters 8 and 9. One of the advantages of field descriptors in QSAR is that fields also allow consideration of structurally diverse data sets. Second, atom-based parameters in QSAR are difficult to automate because molecular modeling programs often renumber atoms during the model-building process. Establishing the required one-to-one correspondence of atoms for QSAR manually is time-consuming. In contrast, the CoMFA field technology circumvents these issues by describing each molecule in terms of a consistent set of grid points and inferred interaction potentials at the grid points. This is because the grid cage definition can be made invariant over all molecules of the data set (thus having an equivalent collection of descriptors). However, the user must define and implement a molecular superposition over all molecules in the data set before beginning any 3-D QSAR study. This requirement is not without challenging difficulties.

E. REPRESENTATIVE STUDIES

To date, two 3-D QSAR studies using the E-State and HE-State fields have been performed. In the first (Kellogg et al., 1996), the steroid test set of Cramer et al. (1988) was examined with steric, electrostatic, and hydropathic (Kellogg et al., 1991) E-State, and HE-State fields. Here, several of the new E-State CoMFA models have better PLS statistics than the steric/electrostatic models. The differences were most striking in models where (i) the E-State field alone produced a model with [cross-validated (predictive) r^2] $q^2 = 0.791$; (ii) adding either the E-State or the HE-State fields to the standard fields improved q^2 by 3–5% over standard CoMFA; and (iii) the combination of the E-State and HE-State fields yielded a CoMFA model with $q^2 = 0.803$. Also interesting was the finding that the E-State fields usually claim a 50% or more contribution when they are combined with other fields. In cases in which the steric, electrostatic, E-State, and HE-State fields were included in single models, the E-State fields claimed 60–70% contribution. This level of relative contribution indicates that the E-State and HE-State 3-D fields provide the bulk of the necessary information to describe the biological performance of the steroid data set.

The ryanoid molecule data set of Welch et al. (1994) was also examined with a multifield CoMFA using steric, electrostatic, HINT (hydropathic), E-State, H-bond, and indicator fields (Kellogg, 1997). Of the many models produced, several had superior statistical metrics to the standard CoMFA results with improvements of 6–10% in q^2 being typical. The combination of the E-State and H-bond fields was shown to be particularly strong in several of the models reported.

For the purposes of drug discovery, the fields should be chosen to maximize the types of information that will yield a lead compound or improve the activity in an existing series. As much as we wish it were otherwise, for ease of computation and comprehension, ligand binding is a concerted process and not a neat sum of individual terms, contacts, and interactions. Our arbitrary terms for describing interactions such as "steric," "electrostatic," "hydrophobic," and "H-bonding" are only a construct. The E-State field, which incorporates several important electronic and steric features into a simple to calculate, easy to visualize, 3-D representation, should be regularly included in 3-D QSAR studies.

III. MOLECULAR AND GROUP POLARITY INDEX

The relative polarity of a molecule may be represented by the position of an encoding index on a scale of polarity defined in some manner. Using the relative scalar position concept from the development of the kappa shape index (Kier, 1987), a general scheme to describe molecular polarity has been developed. A definition of polarity is constructed around the collective electrotopological accessibility of atoms in a molecule. This is reckoned by summing the intrinsic states of atoms in a molecule, Sum I_i, over the whole molecule. This value is assumed to lie between the extremes of minimal polarity of an isoconnective alkane, Sum I_0, and the maximal polarity, Sum I_{max}, for the same structural framework:

$$\text{Sum } I_0 \geq \text{Sum } I_i \geq \text{Sum } I_{max} \qquad (4.5)$$

The Sum I_0 value is derived from the skeleton structure of molecule i in which each nonhydrogen atom is replaced with a carbon sp^3 atom forming an isoconnective alkane as a reference structure. For example, the alkane skeleton of acetone is isobutane. This is considered to epitomize the most nonpolar structure model of molecule i. The maximum polarity for molecule i is based on a more arbitrary choice of derivatives. This hypothetical molecule must be rich in electronegative atoms. One could conceive of a molecule bristling with fluorine atoms or other polar groups, retaining the same topological structure as the subject molecule i. The choices here are daunting, particularly in a large molecule. A reasonable compromise is to adopt a large number relevant to the molecule for the term, Sum I_{max}. A provisional choice is to use a number related to the number of atoms, c, in molecule i. The relationship Sum $I_{max} = c^2$ was adopted to encode this reference structure.

To develop a polarity index from these ingredients, the assumption is made that the polarity of molecule i, expressed as Sum I_i, is related to the two extreme values as two ratios:

$$\frac{\text{Sum } I_0}{\text{Sum } I_i} \quad \text{and} \quad \frac{\text{Sum } I_{max}}{\text{Sum } I_i} = \frac{c^2}{\text{Sum } I_{max}} \tag{4.6}$$

The product of these two ratios yields a single number, Q:

$$Q = \frac{\text{Sum } I_0 \times c^2}{(\text{Sum } I_i)^2} \tag{4.7}$$

The Q index is a descriptor encoding attributes that may be likened to molecular polarity. A test of this index was made by relating these values to an appropriate physical property such as the partition coefficient (Table 4.3). The correlation with log P is given as follows:

$$\log P = 0.80Q - 0.37$$

$$r^2 = 0.89, \; s = 0.23, \; n = 21$$

TABLE 4.3 Q Values and Partition Coefficients for Different Compounds

No.	Compound	Q	Log P
1	Butanol	1.36	0.88
2	Isobutanol	1.34	0.61
3	t-Butanol	1.32	0.37
4	3-Pentanol	1.81	1.14
5	Cyclohexanol	2.41	1.23
6	Benzyl alcohol	2.15	1.10
7	Methyl butyl ether	2.50	1.53
8	Propyl isopropyl ether	3.03	1.83
9	Diethyl ether	1.93	1.03
10	Dipropyl ether	3.09	2.03
11	Ethyl acetate	1.19	0.73
12	Isopropul acetate	1.57	1.03
13	Butyl acetate	2.00	1.73
14	sec-Butyl acetate	1.98	1.53
15	Propanoic acid	0.78	0.25
16	Phenylacetic acid	2.02	1.41
17	Diphenylacetic acid	4.03	3.05
18	Propylamine	1.38	0.48
19	Hexylamine	3.09	1.98
20	Butanone	1.10	0.29
21	2-Methyl pentan-3-one	1.50	1.09

TABLE 4.4 Q Values and Partition Coefficients of
Neutral Amino Acids

No.	Amino acid	π Value	Q
1	Serine	−0.04	0.25
2	Threonine	0.26	0.55
3	Alanine	0.31	0.50
4	Valine	1.22	1.69
5	Tyrosine	0.96	2.20
6	Methionine	1.23	2.23
7	Isoleucine	1.80	2.34
8	Leucine	1.70	2.34
9	Phenylalanine	1.79	2.92
10	Tryptophane	2.25	4.02

A comparison with partition coefficients of neutral amino acid side chains is shown in Table 4.4. It appears that this index has some potential for capturing useful information about molecular and group polarity.

IV. THE ATOM-TYPE INDICES

We have introduced an extension of the E-State called the atom-type indices, similar to group additive schemes, in which an index appears for each atom type in the molecule together with its contribution based on the E-State index (Hall and Kier, 1995). It is possible that a limited number of atom-type indices may be used for a given application, especially for biological data in which only a few atom types are required for a quality QSAR equation. For some biological QSAR models, a type of skeletal superposition is used so that the E-State values for corresponding atoms may be entered as variables in regression analyses. The development of the atom-type E-State description provides the basis for application to a wider range of problems in which the E-State formalism is applicable without the need for superposition or a closely related set of structures.

A. ATOM-TYPE CLASSIFICATION SCHEME

In this classification scheme, each atom in the molecule is identified by its valence state, including the number of attached hydrogen atoms. Classification is based on four hydride group characteristics: (i) atom (element) identification;

(ii) valence state, including aromaticity indication; (iii) the number of bonded hydrogen atoms; and (iv) in a few cases, the identity of other bonded atoms.

1. Atom (Element) Identity

This characteristic is directly represented by the atomic number, Z, which is an unambiguous element identifier.

2. Valence State

Valence state identification is less direct than element identification by atomic number. Valence state is usually based on the number of valence electrons assigned to sigma, pi, and lone-pair orbitals. Also included is a representation of resonance forms and/or aromaticity. It has been shown that the molecular connectivity delta values, δ^v and δ, are useful in the designation of valence state (Kier and Hall, 1981). For encoding a particular valence state, the delta value sum, $\delta^v + \delta$, is used as shown in Table 4.5. The delta value summation, $\delta^v + \delta$, is called the valence state indicator (VSI). The valence state designation also makes use of a marker for an atom which is part of an aromatic system as shown in Table 4.5. In this case, we use a simple bivariate indicator: AR = 1 for aromatic and 0 for nonaromatic. This indicator may be directly supplied, as in the case of lowercase letters in SMILES strings, or determined from the Huckel rule.

3. Number of Bonded Hydrogen Atoms

It has proven useful in various group additive schemes to create separate classes for atoms in the same valence state but with a different number of hydrogen atoms, for example, —CH₃, —CH₂—, —CH<, and >C<; —NH₂, —NH—, and —N<; and —OH and —O—. Hydrogen atoms are important because of their role in intermolecular interactions and their contribution to physical properties. Therefore, it is necessary to consider each of these hydride groups as a different class. Examination of Table 4.5 shows that the VSI, $\delta^v + \delta$, classifies these groups appropriately in addition to valence state designation (with a few exceptions as described later).

4. Distinguishing Identity of Bonded Atoms

There are a few cases in which the atomic number and the valence state designation do not distinguish among atoms that clearly belong to different groups based on chemical experience. The following groups require bonded-atom analysis:

1. Allenic and acetylinic carbon atoms
2. A carbon atom at the juncture of two aromatic rings as in the 9 and 10 positions in naphthalene and an aromatic carbon which is bonded to a substituent, such as the ipso atom in 1-naphthol
3. The tertiary amine nitrogen, the nitro nitrogen, and the nitrogen in a pyridine N-oxide
4. The sulfur in a sulfide and in a disulfide link

In these four cases, use is made of the adjacency matrix to determine the nature of bonded atoms as a basis for assignment to the appropriate group. For example, the 9 and 10 carbons in naphthalene are bonded to three atoms with aromatic markers, whereas the aromatic carbon atom with an external single bond is bonded to only two atoms with aromatic markers.

This classification scheme may be viewed as a 3-D array with the atomic number as one dimension, the VSI, ($\delta^v + \delta$), as the second dimension, and the aromatic indicator as the third dimension. This array correctly classifies all the 52 groups listed in Table 4.5. The VSI classifies 42 groups. Six more are properly classified by the bonded-atom analysis described previously, and they are marked with a single asterisk in Table 4.5. The remaining 4 cases are resolved by use of the delta difference, $\delta^v - \delta$, as shown in Table 4.5 in which they are marked with the letter j. For example, for the tertiary nitrogen atom, $>N-$, $\delta^v - \delta = 4$, whereas for $=N-$, $\delta^v - \delta = 3$. In each of these cases, the delta sum is the same for both groups but the delta difference is not the same. The count of each atom type is accumulated along with the sum of E-State values for all groups of the same type. These two sets of indices are stored and made available for the user. In principle, this scheme can be extended to other elements in their various hydride groups (and this currently is extended in Molconn-Z for 80 atom types).

Table 4.5 provides a list of atom types along with the chemical symbols used. For example, in the symbol $S^T(-CH_2-)$, S^T is the sum of E-State values for all the $-CH_2-$ groups in the molecule and $-CH_2-$ represents the formula for the hydride group. In this manner, it is possible to distinguish between $-CH_2-$ and $=CH_2$. Also, $S^T(...CH...)$ is the sum of E-State indices for the CH in an aromatic ring and $S^T(-OH)$ is the sum of E-State indices for $-OH$ groups in the molecule. [Note: In some of our papers (Hall and Kier, 1995; Hall et al., 1995; Hall and Story, 1995; Hall and Vaughn, 1997) a different set of symbols was used; for example, SaasC rather than $S^T(\overset{|}{\underset{C}{\ldots}})$ and SdssC rather than $S^T(=C<)$. Symbols containing special characters such as "—," "=," or "<" are not permitted in computer programming, where these symbols originated.]

TABLE 4.5 Atom Types and Related Information for Atom-Type E-State Indices

No.	Atom group[a]	Valence state indicator					Group symbol[h]	Chemical symbol[i]
		Z^b	δ^{vc}	δ^d	$\delta^v + \delta^e$ $\delta^v - \delta^f$		AR^g		
1	—CH$_3$	6	1	1	2	0	0	sCH3	—CH$_3$
2	=CH$_2$	6	2	1	3	1	0	dCH2	=CH$_2$
3	—CH$_2$—	6	2	2	4	0^j	0	ssCH2	—CH$_2$—
4	≡CH	6	3	1	4	2^j	0	tCH	≡CH
5	=CH—	6	3	2	5	1	0	dsCH	=CH—
6	aCHa	6	3	2	5	1	1	aaCH	...CH...
7	>CH—	6	3	3	6	0^j	0	sssCH	>CH—
8	=C=	6	4	2	6	2	0	ddC^k	=C=
9	#C—	6	4	2	6	2	0	tsC^k	≡C—
10	=C<	6	4	3	7	1	0	dssC	=C<
11	aCa—	6	4	3	7	1	1	aasC^k	..C..
12	aaCa	6	4	3	7	1	1	aaaC^k	..C..
13	>C<	6	4	4	8	0	0	ssssC	>C<
14	—NH$_3$[+1]	7	2	1	3	1	0	sNH3	—NH$_3^+$
15	—NH$_2$	7	3	1	4	2	0	sNH2	—NH$_2$
16	—NH$_2$—[+1]	7	3	2	5	1	0	ssNH2	—NH$_2^+$—
17	=NH	7	4	1	5	3	0	dNH	=NH
18	—NH—	7	4	2	6	2	0	ssNH	—NH—
19	aNHa	7	4	2	6	2	1	aaNH	...NH...
20	#N	7	5	1	6	4^j	0	tN	≡N
21	>NH—[+1]	7	4	3	7	1	0	sssNH	>N$^+$—
22	=N—	7	5	2	7	3	0	dsN	=N—
23	aNa	7	5	2	7	3	1	aaN	...N...
24	>N—	7	5	3	8	2	0	sssN^k	>N—
25	—N≪	7	5	3	8	2	0	ddsN (nitro)^k	—N=
26	aaNs	7	5	3	8	2	1	aasN (N oxide)^k	..N..
27	>N<[+1]	7	5	4	9	−1	0	ssssN (onium)	>N$_2^+$<
28	—OH	8	5	1	6	4	0	sOH	—OH
29	=O	8	6	1	7	5	0	dO	=O
30	—O—	8	6	2	8	4	0	ssO	—O—
31	aOa	8	6	2	8	4	1	aaO	...O...
32	—F	9	7	1	8	6	0	sF	—F

TABLE 4.5 *(continued)*

No.	Atom group[a]	Valence state indicator						Group symbol[h]	Chemical symbol[i]
		Z^b	$\delta^{v\,c}$	δ^d	$\delta^v + \delta^e$	$\delta^v - \delta^f$	AR[g]		
33	—PH₂	15	3	1	4	2	0	sPH2	—PH₂
34	—PH—	15	4	2	6	2	0	ssPH	—PH—
35	>P—	15	5	3	8	2	0	sssP	>P—
36	—>P=	15	5	4	9	1	0	dsssP	—P=
37	—>P<	15	5	5	10	0	0	sssssP	—P
38	—SH	16	5	1	6	4	0	sSH	—SH
39	=S	16	6	1	7	5	0	dS	=S
40	—S—	16	6	2	8	4	0	ssS[k]	—S—
41	aSa	16	6	2	8	4	1	aaS	...S...
42	>S=	16	6	3	9	3	0	dssS (sulfone)	>S=
43	==>S=	16	6	4	10	2	0	ddssS (sulfate)	—S—
44	—>S<—	16	6	6	12	0	0	ssssssS	S
45	—Cl	17	7	1	8	6	0	sCl	—Cl
46	—SeH	34	5	1	6	4	0	sSeH	—SeH
47	=Se	34	6	1	7	5	0	dSe	=Se
48	—Se—	34	6	2	8	4	0	ssSe	—Se—
49	>Se=	34	6	3	9	3	0	dssSe	—Se—
50	>Se=	34	6	4	10	2	0	ddssSe	—Se—
51	—Br	35	7	1	8	6	0	sBr	—Br
52	—I	53	7	1	8	6	0	sI	—I

[a] Indication of atom groups, according to valence state, together with number of bonded hydrogens. —, single; =, double; #, triple; a, aromatic; $\ll$, two double bonds or two resonance single/double bonds (nitro group); —>, three single bonds.

[b] Atomic number of element.

[c] Valence connectivity delta value.

[d] Simple connectivity delta value.

[e] Sum of valence and simple delta value.

[f] Difference of valence and simple delta value.

[g] Indicator that atom is part of aromatic system: 1 indicates aromaticity.

[h] Symbols for atom type as used in conjunction with the atom-type E-State index, specially designed for use in computer programs for statistical analysis. No special symbols (e.g., "—" and "=") are used.

[i] Standard chemical symbol.

[j] Cases in which the delta difference, $\delta^v - \delta$, is required as an additional classification criterion (see text).

[k] Denotes an atom which can be distinguished from another only by analysis of neighboring atoms (see text).

TABLE 4.6   Computed E-State Values for Dopamine
and the Summation Atom-Type E-State Indices

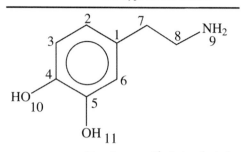

Atom No.	Atom-type symbol	Electrotopological State index value
1	...C...	0.933
2	...CH...	1.738
3	...CH...	1.459
4	...CH...	−0.092
5	...C...	−0.087
6	...C...	1.516
7	—CH$_2$—	0.716
8	—CH$_2$—	0.546
9	—NH$_2$	5.305
10	—OH	8.931
11	—OH	9.036

Atom-type E-State symbol	Sum index value
S^T(—OH)	17.967
S^T(—NH$_2$)	5.305
S^T(...CH...)	4.713
S^T(...C...)	1.489
S^T(—CH$_2$—)	1.262

B. SAMPLE CALCULATION

Table 4.6 gives the E-State atom values as well as the sum of atom-type E-State values for dopamine, for example. The individual E-State values (atom level) are given in the top of Table 4.6. Values range from −0.087 for a substituted phenyl ring carbon atom up to 9.036 for one of the hydroxyl groups. Note that the two —OH groups do not have identical E-State values because they are not chemically identical in the structure. The atom-type E-State values are given in the bottom of Table 4.6. There are five values corresponding to the five atom types as classified by Molconn-Z.

C. Significance of the Atom-Type E-State Values

The E-State indices have demonstrated considerable usefulness in the establishment of QSAR equations and as chemical structure spaces. The ability to focus on a small number of atoms is a powerful advantage in their application. The fact that they encode important electronic and topological information endows them with the ability to portray significant pharmacological information for database characterization. The addition of atom typing extends the usefulness of the E-State indices so that large, diverse data sets may be examined. Recent studies attest to the potential of this addition to the E-State methodology (Hall *et al.*, 1995; Hall and Story, 1995; Hall and Vaughn, 1997; Kier and Hall, 1992). The atom-by-atom matchup (or topological superposition) required when individual E-State values are used and which is so useful for biological studies is not useful for general physicochemical properties, such as aqueous solubility and boiling point, or for large heterogeneous data sets. Atom groups do not present themselves to each other in a strict atom-by-atom matching arrangement; rather, molecules present atom features to each other in the liquid state in a random fashion. The groups on the molecular surface make significant contributions along with those features that are polar or that participate in hydrogen bonding.

The atom-type E-State scheme makes possible an important combination of structure information for QSAR analysis and for similarity and diversity analyses. The atom-type E-State indices combine three very important aspects of structure information:

1. The electron accessibility associated with each atom type, characteristic of the E-State index, as described in Chapters 2 and 3
2. An indication of the presence or absence of a given atom type
3. A count of the number of atoms of a given atom type

These structure attributes are combined in the atom-type E-State formalism as in no other form of structure representation.

D. Atom-Type Hydrogen E-State Indices

The hydrogen atoms in organic structures may also be classified. Two general types are commonly used: polar and nonpolar. We selected the following hydride groups as having polar X–H bonds in the generally decreasing order of polarity: —OH, =NH, —NH$_2$, —SH, —NH—, and ≡CH. Hydrogen atoms which are part of other hydride groups are considered nonpolar. On this basis, atom-type hydrogen E-State indices are computed and symbolized in a fashion similar to that for the atom type described previously. For example, for the —OH group the symbol is SH^T(—OH), where the S is the sum of the hydrogen E-State values for all —OH groups in the molecule (the average value could also be used), and the H indicates hydrogen E-State. For the imine group, =NH, the sum atom-type symbol is SH^T(=NH).

As will be illustrated in later QSAR examples, sometimes it is useful to group some of the values together. SH^T(other) is the sum of hydrogen E-State values for all nonpolar hydrogen atoms. SH^{T}Csat represents the hydrogen atoms in saturated carbon groups (—CH$_3$, —CH$_2$—, and >CH—): SH^T(Csat) = SH^T(—CH$_3$) + SH^T(—CH$_2$—) + SH^T(>CH—).

V. SUMMARY

The richness of the E-State formalism has been illustrated in this chapter by discussing several useful extensions of the method. Some of these extensions, such as the HE-State, have already been used successfully in QSAR and will be described in subsequent chapters. The atom-type E-State is now widely used. Others, such as the polarity index Q, are currently under investigation.

VI. A LOOK AHEAD

With the E-State formalism described along with its extensions, it is appropriate that we investigate strategies for its implementation. Chapter 5 presents several general approaches to the use of the E-State and the implications of these strategies. The remaining chapters present specific studies which have used the E-State in various situations.

REFERENCES

Cramer, R. D., Paterson, D. E., and Bunce, J. D. (1988). Comparative molecular field analysis (CoMFA). Effect of shape on binding of steroids to carrier proteins. *J. Am. Chem. Soc.* **110**, 5959–5967.

Glennon, R. A., Herndon, J. L., and Dukat, M. (1994). Epibatidine-aided studies toward definition of a nicotine receptor pharmacophore. *Med. Chem. Res.* **4**, 461–473.

Gough, J., and Hall, L. H. (1999). Modeling the toxicity of amide herbicides using the electrotopological state. *Environ. Toxicol. Chem.* (in press).

Hall, L. H., and Kier, L. B. (1995). Electrotopological state indices for atom types. *J. Chem. Inf. Comput. Sci.* 35, 1039–1045.

Hall, L. H., and Story, C. T. (1995). Boiling point and critical temperature of a heterogeneous data set: QSAR with atom-type E-State indices using artificial neural networks. *J. Chem. Inf. Comput. Sci.* 36, 1004–1014.

Hall, L. H., and Vaughn, T. A. (1997). QSAR of phenol toxicity using E-State and kappa shape indices. *Med. Chem. Res.* 7, 407–416.

Hall, L. H., Kier, L. B., and Brown, B. B. (1995). Molecular similarity based on novel atom-type E-State indices. *J. Chem. Inf. Comput. Sci.* 35, 1074–1080.

Hocart, S. J., Reddy, V., Murphy, W. A., and Cody, D. H. (1993). 3-D QSAR of somatostatin analogues I. CoMFA of growth hormones release-inhibiting potencies. *J. Med. Chem.* 38, 1974–1089.

Jhon, M. S., and Sheraga, H. (1988). Analytic intermolecular potential functions from ab initio SCF calculations of interaction entropies between CH_4, CH_3OH, CH_3COOH, CH_3COO^- and water. *J. Phys. Chem.* 92, 7216–7225.

Kellogg, G. E. (1997). Finding optimum field models for 3-D CoMFA. *Med. Chem. Res.* 7, 417–427.

Kellogg, G. E., Semus, S. F., and Abraham, D. J. (1991). HINT—A new method of empirical hydrophobic field calculation for CoMFA. *J. Comput.-Aided Mol. Design* 5, 545–553.

Kellogg, G. E., Kier, L. B., Gaillard, P., and Hall, L. H. (1996). E-State fields: Application to 3-D QSAR. *J. Comput.-Aided Mol. Design* 10, 513–520.

Kier, L. B. (1987). Indexes of molecular shape from chemical graphs. *Med. Res. Rev.* 7, 417–426.

Kier, L. B., and Hall, L. H. (1981). Derivation and significance of valence molecular connectivity. *J. Pharm. Sci.* 70, 583–589.

Kier, L. B., and Hall, L. H. (1992). Atom description in QSAR models: Development and use of an atom level index. *Adv. Drug Res.* 22, 1–35.

Kroemer, R. T., Ettmayer, P., and Hecht, P. (1995). 3-D QSAR of human immunodeficiency virus type-1 proteinase inhibitors: CoMFA of 2-heterosubstituted statine derivatives—Implications for the design of novel inhibitors. *J. Med. Chem.* 38, 4917–4928.

Waller, C. L., and Kellogg, G. E. (1996). Adding chemical information to CoMFA models with alternative 3-D QSAR fields. *Netscience* 2(1).

Wee, S. S., Kim, S., Jhon, M. S., and Sheraga, H. (1990). Analytical intermolecular potential functions from ab initio self-consistent field calculations for hydration of methylamine and methylammonium ion. *J. Phys. Chem.* 94, 1656–1662.

Welch, W., Ahmad, S., Gerzon, K., Humerickhouse, R. A., Besch, H. R., Ruest, L., Deslongchamps, P., and Sutko, J. L. (1994). Structural determinants of high-affinity binding of ryanodine receptor: A CoMFA analysis. *Biochemistry* 33, 6074–6085.

CHAPTER 5

Strategies for Use of the E-State

I. EXPERIMENTAL DESIGN

The E-State indices form a pattern within the chemical graph representing the structure of a molecule. If we alter that molecular structure by introducing or changing a substituent, a corresponding change occurs in all E-State values within each new molecule, including the atoms in the unaltered portion. If a series of such molecules is exposed to a potential binding site on a protein, we anticipate that one or a few positions in the unaltered part of the molecule may be influential in determining the binding affinity. This response is the manifestation of a pharmacophore. The question we may then ask is whether the relative binding affinity within the compound series corresponds or correlates with any E-State values associated with atoms in the compounds or with the influence of these atoms on another part of a molecule.

When the series of molecules is varied in only one position, all E-State values change as a direct function of the intrinsic state, I, of the changing substituent. Such changes are shown in Table 5.1 in which increasing $I(X)$ values produce monotonic changes in every atom E-State value. The degree of change and the sign of the change of an E-State value in the molecule is a function of that particular atom's I value and its relationship to the I value of the changing substituent. Tables 5.1–5.3 show small databases of series of substituents X, Y, and Z, with intrinsic state values ranging from 2 to 8, for the fragment $—CH_2(C=O)CH_2NH_2$. With only one substituent on the molecule (Table 5.1), the relative change in the E-State values experienced at each atom through the series is constant. As a result, there is a limited variety of information in this data set. The correlation matrix in Table 5.4 for the structure data in Table 5.1 shows significant intercorrelations. As a consequence, if we were to seek a correlation between the E-State values on one atom and a measured molecular property, most of the five atom positions would show similar levels of correlation. We are confined to limited information when exploring only one variable. Indeed, principal component analysis shows only one independent piece of information.

If we introduce a second substituent in another position, such as a substituent on the amine group shown in Table 5.2, the intercorrelation among the five atoms is reduced, as shown in the correlation matrix in Table 5.5 for the

TABLE 5.1   Substituent Influence on E-State Values

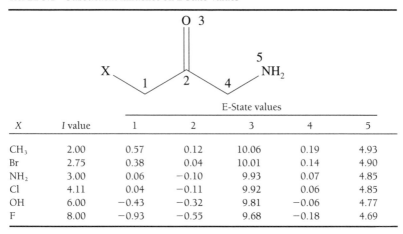

X	I value	1	2	3	4	5
				E-State values		
CH₃	2.00	0.57	0.12	10.06	0.19	4.93
Br	2.75	0.38	0.04	10.01	0.14	4.90
NH₂	3.00	0.06	−0.10	9.93	0.07	4.85
Cl	4.11	0.04	−0.11	9.92	0.06	4.85
OH	6.00	−0.43	−0.32	9.81	−0.06	4.77
F	8.00	−0.93	−0.55	9.68	−0.18	4.69

structure data in Table 5.2. Several pairs of atoms show a degree of independence and can be used in a two-variable equation analyzing some measured molecular property. Principal component analysis reveals two independent pieces of information. The introduction of a third substituent in the molecules, as shown in Table 5.3, produces a greater decline in the intercorrelation among

TABLE 5.2   Data Set B

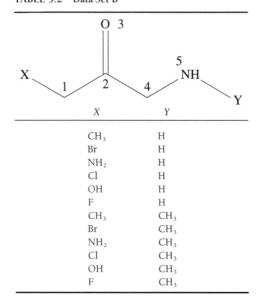

X	Y
CH₃	H
Br	H
NH₂	H
Cl	H
OH	H
F	H
CH₃	CH₃
Br	CH₃
NH₂	CH₃
Cl	CH₃
OH	CH₃
F	CH₃

TABLE 5.3   Data Set C

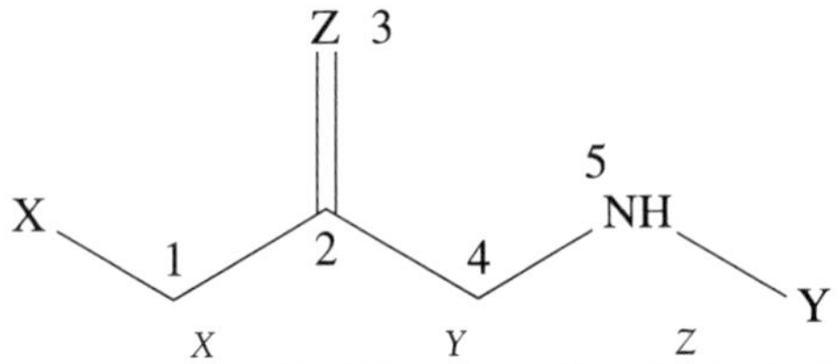

Z 3

5
NH

X

1   2   4   Y

X	Y	Z
CH$_3$	H	O
Br	H	O
NH$_2$	H	O
Cl	H	O
OH	H	O
F	H	O
CH$_3$	CH$_3$	O
Br	CH$_3$	O
NH$_2$	CH$_3$	O
Cl	CH$_3$	O
OH	CH$_3$	O
F	CH$_3$	O
CH$_3$	H	S
Br	H	S
NH$_2$	H	S
Cl	H	S
OH	H	S
F	H	S

TABLE 5.4   Correlation Matrix for Data in Table 5.1:
Single Position of Substitution

	Position			
	2	3	4	5
1	0.99	0.58	0.96	0.82
2		0.54	0.99	0.88
3			0.44	0.93
4				0.73

TABLE 5.5   Correlation Matrix for Data in Table 5.2:
Two Positions of Substitution

	Position			
	2	3	4	5
1	0.97	0.70	0.67	0.01
2		0.85	0.83	0.24
3			0.99	0.71
4				0.74

TABLE 5.6   Correlation Matrix for Data in Table 5.3:
Three Positions of Substitution

	Position			
	2	3	4	5
1	0.76	0.25	0.76	0.19
2		0.81	0.81	0.40
3			0.42	0.60
4				0.20

the five designated atoms, as is shown in the correlation matrix in Table 5.6 for the structure data in Table 5.3. It is clear that substitution in the three separate positions produces a more diverse set of structures.

From these analyses, it is clear that molecular series with multiple substitution patterns represent the best opportunity to identify one or more positions within a molecule as being significantly correlated with a molecular property. In a sense, the distinctive electrotopological attribute of an atom is more fully developed when it is under the influence of several structure changes operating along different primary paths. Consequently, a well-designed series of compounds with this in mind will contain multiple structure changes on some basic fragment.

II. PRELIMINARY DATA ANALYSIS

In the first stages of a quantitative structure–activity relationship (QSAR) analysis, it is often useful to examine the data set to determine whether all the information is meaningful for that particular data set. A few strategies have been widely used and will be mentioned briefly here.

A. THE RANGE OF VARIABLES: VARIANCE

With the use of currently available software, there are many structure descriptor variables which may be computed and made available for the analysis. There may be 10–15 atom-level E-State indices. Furthermore, there may be 20–30 atom-type E-State indices. In addition, there may be several whole molecule indices which represent overall shape and skeletal branching. For small data sets the actual number may not be large, but for the larger sets becoming available from combinatorial chemistry the number of useful variables may be quite large.

It is important to determine the numerical range within the data set. Any statistical package can be used to compute these values, including variance, for each variable. Those with limited range possessing small variance can be eliminated. It may be useful to examine those with small variance to determine the reason for this condition. For example, a narrow range may arise simply from very little variation in a variable from compound to compound because of the homogeneity of the structures. On the other hand, a narrow range (small variance) may arise because only a limited number of compounds have a nonzero value for a variable. This may be an important clue, indicating a subset of compounds which is structurally different. Such a subset may represent a cluster in the data or a rare feature which may or may not be significant in the QSAR analysis. It is useful to be aware of such cases during the course of the analysis.

B. Independence of Variables

A second aspect of the preliminary analysis of any set of structure variables is that the variables are not necessarily independent. That is, there may be significant intercorrelations among the variables. Generally, intercorrelations should be eliminated or at least reduced by deletion of some of the variables. Most statistical packages make available the correlation matrix of the variables. For example, the SAS procedure for principal components produces the variances and the correlation matrix in addition to the principal component eigenvalues and eigenvectors. An inspection of the correlation matrix reveals those pairs of variables with significant intercorrelations, for example, with $r^2 > 0.80$. For such cases, one of the pair of variables should be deleted from the data set. That is, the data matrix should be reduced. Not only does such deletion reduce subsequent computational time but also it avoids certain computational problems which arise with intercorrelated variables.

It is a matter of experience to make judicious decisions as to which one of a pair of variables to eliminate. The QSAR analyst must rely on past experience or advice from more experienced investigators. A list of the variables eliminated and their paired correlates should be kept. When QSAR model equations are obtained later, one can determine whether any of the paired correlates were actually found to be important. Then one can try substitution of the correlate to determine whether an improved model equation can be obtained.

C. Eigenvalues and Principal Components

Another aspect of this initial analysis is inspection of the principal component eigenvalues. The accumulated eigenvalues indicate the fraction of data variance which is included or spanned by the set of principal components. The number of components required to span 95–98% of the data variance is an indication of the number of independent pieces of information contained in the set of variables. This number represents the upper limit of the number of variables which may be reliably included in a model.

D. DATA PLOTS: ACTIVITY AND STRUCTURE DESCRIPTORS

The final aspect of the initial analysis is the plotting of the activity versus many of the independent structure variables. In these plots one can look for variables which are related to activity, for example, which yield a linear plot. Furthermore, sometimes clustered data are revealed, and sometimes a significant outlier may be spotted. Nonlinear relationships may begin to reveal themselves in such plots. In any case, one begins to develop a useful feel for the data.

E. OVERALL PERSPECTIVE

All these approaches tend to keep the data set from becoming a simple black box to which one applies other black box statistical methods. In the final analysis the QSAR model and its attendant statistical results should reveal the basis for any relationship found between the molecular structure and the properties under investigation. The E-State indices represent molecular structure in an intuitive way as well as the formal presentation in Chapter 2. The E-State indices lend themselves very well to careful analysis which uses many methods, as will be illustrated in subsequent chapters.

III. TOPOLOGICAL SUPERPOSITION STRATEGY

In many data sets used for structure–activity analyses, all the molecules share a common molecular skeleton; that is, a scaffold with substituents occurring at various positions. The substituents vary, producing varying atomic character in the common skeletal atoms. In these data sets the atoms in the scaffold should be numbered in a common manner in each molecule. The E-State index values for each atomic position in the skeleton may then be used as variables in structure–activity analyses. For example, if the parent molecule is an m-chlorophenol, then there are eight atoms—one phenolic OH, six aromatic carbons, and one chlorine—in the structurally invariant skeleton. It is observed that, although a given atom may be located in the same position in the skeleton, the computed E-State value does vary with the variation in the substituents. This E-State value of an atom at such a position encodes the electron accessibility at that atom. It is the objective of a structure–activity analysis to determine whether a particular position is one of a profile of E-State values whose variation parallels the experimental values of a property under investigation.

In some cases a specific position in the common skeleton may be occupied by atoms of different elements in the different compounds in the data set. An example will be provided in Section IX in which one specific skeletal position is occupied by C, N, or O.

This approach to structure–activity analysis is based on a topological super-position principle. That is, atoms in similar positions within a molecular scaffold may play similar roles in the interactions which give rise to the value of a property. Within this principle it is not necessary to deal with the more complicated needs of a three-dimensional superposition. In this chapter, we examine cases in which only atom-level E-State indices are required for a quality structure–activity model. Other chapters present cases in which atom-type E-State indices as well as other topological indices are used.

IV. DATA SELECTION AND MANIPULATION

In anticipation of the use of E-State indices to explore QSAR models, it is worthwhile to digress from the structure side of structure–activity equations and discuss some aspects of data selection and manipulation. For purposes of describing some problems in this arena, a hypothetical set of data has been created (Table 5.7). In the following sections, we discuss problems in using literature or report data.

A. FORM OF ACTIVITY QUANTITATION

The activity in the series in Table 5.7 is expressed in mg/kg. There is enough difference in molecular weights among the compounds to render this an undesirable metric. The activity should be expressed on a molar basis since molecular structure, and not mass, is being compared to biological activity or a physical property.

B. INDEFINITE ACTIVITY EXPRESSION

Molecule 5 in the data set has an activity expressed as >2500 mg/kg, whereas the other data are expressed using definite values. This indefinite activity value should be discarded. We cannot assume that the activity is close to

TABLE 5.7 Data Set Illustrating Problems

Observed	R_1	R_2	ED_{50} (mg/kg)	Parameter X
1	CH_3O	CH_3	15.3	3.8
2	OH	H	2.6	1.7
3	Cl	CH_3	22.8	4.8
4	CH_2COOH	CH_3CH_2	0.5	0.3
5	Br	CH_3	>2500	6.2
6	C_6H_5	H	0.2	3.4
7	CH_3CH_2	CH_3	110.0^a	9.7
8	CH_3O	H	2400	15.2
9	$CH(CH_3)_2$	CH_3	500	20.4
10	CH_3	C_6H_5	40.6^b	16.1
11	NO_2	H	0.1	0.7
12	$ClCH_2$	$CH(CH_3)_2$	1500	13.3
13	CH_3CH_2	$(+)\,CH_3CH(OH)CH_2$	500	8.2
14	CH_3CH_2	$(-)\,CH_3CH(OH)CH_2$	50	8.2
15	F	CH_3	0.04	1.3
16	F_3C	CH_3CH_2	0.08	1.4^c
17	CH_3CH_2	$CH_3CH_2CH_2$	1.5	2.4
18	$-COCl$	H	3.5	1.9
19	C_6H_5	CH_3	18.6	11.6
20	Cl	CH_3CH_2	20.4	6.4

[a] Data from Jones et al. (1965).
[b] Low solubility; tests run with 50% diglyme.
[c] Value estimated from fragment method.

2500 mg/kg. On some occasions an activity expression of <0.001 or a small number relative to much larger numbers in the data set may be considered to be zero. Caution is advised in these situations.

C. UNCERTAIN DEGREE OF IONIZATION

Molecule 4 in the data set is a carboxylic acid. Caution is advised here because the degree of ionization may be an important factor in the properties of the molecule. It may be necessary to discard it if the uncertainty jeopardizes the accurate consideration of the molecular structure.

D. DATA FROM SEVERAL SOURCES

Molecule 7 in the data is obtained from another source. There is no evident assurance that the same protocol was used in evaluating this compound. It is prudent to consider discarding this entry unless a correspondence with other protocols and procedures is confirmed.

E. Departure from a Protocol

Molecule 10 was studied using a modification of the procedure. These data are accordingly suspect and should probably be discarded in creating a structure–activity model.

F. Stereoisomers

Compounds 13 and 14 are stereoisomers. The activities reported differ 10-fold. The problem here is the encoding of either isomer. At this time, this cannot be done using either a physical property or topological descriptors. The data for compounds 13 and 14 should not be used.

G. Aberrant Parameters

Compound 16 has associated with it an X parameter that was estimated apparently by a method differing from that of the other parameters. A judgment must be made as to the correspondence of this alternative method with the remainder of the parameters. If there is any uncertainty, it should be discarded.

H. Unstable Molecules

As strange as it may seem, some very unstable molecules, such as No. 18 in the series, do appear in some databases. Obviously, this acid chloride would decompose to the corresponding acid when it encountered any water. Therefore, the data for this molecule cannot be used.

I. Use of Log Activities

It is customary to convert these activities to the log values for comparison with structure parameters because of the assumed linear relation between biological response and log(dose) in the midregion of the log(dose)–response curve.

V. STRUCTURE DATA INPUT

In a series of molecules with a portion of the skeleton unchanged (a scaffold), it is essential that a common atom numbering scheme be devised for those atoms in the common skeleton. Consider a series of molecules with the basic structure as shown in Scheme I. In this series, the quinoline ring is common among all the molecules. The ethyl side chain is also found on each molecule. The variable fragments are X, Y, and Z. If the molecules in this series are entered into a program using the SMILES scheme, the numbering will be different for each molecule. The integrity of the common numbering as shown in Scheme I will not be preserved. It is necessary to enter the structure data into the program which computes the E-State values in such a way that the numbering pattern in Scheme I is preserved. If a substituent atom actually occupies one of the sites X, Y, or Z in all the molecules, then that site may be added to the numbering scheme.

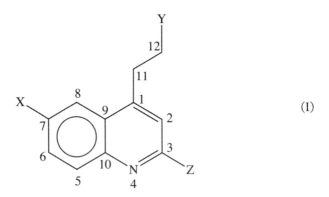

(I)

There may be several ways in which this consistent numbering pattern may be accomplished. Provision is made in the Molconn-Z program for the assignment of numbering in any desired order. These methods have been included in the CD provided with this book. Input formats are described in the Appendix.

VI. CALCULATION OF TAUTOMERS AND EQUIVALENT ATOMS

Occasionally, a molecule may be encountered in which a tautomeric equilibrium is present. Two examples are nitroso-oxime (Scheme II) and an imidazole ring (Scheme III). These structures may be assumed to exist as shown in Schemes II and III and so the assignments of I values for the oxygen and the nitrogen atoms could be as follows:

$$I(>CH-) = 1.33 \qquad I(-OH) = 6.00$$

$$I(>C=) = 1.66 \qquad I(=O) = 7.00$$

$$R_2CH-N=O \quad \rightleftharpoons \quad R_2C=N-OH \qquad (II)$$

$$(III)$$

For Scheme III:
$$I(=N-) = 3.00, I(-NH-) = 2.50$$

There may arise in one of these studies the need to consider an average structure of these atoms in which the tautomer ratio is close to one. Other considerations may impel an investigator to treat the average structure of these molecules. In these cases, an average value of the oxygen in Scheme II and nitrogen in Scheme III is assigned:

$$I(O) = 6.50, \quad I(N) = 2.75$$

Other examples are the imine–enamine pair (Scheme IV), the keto–enol tautomers (Scheme V), and the lactam–lactim tautomers (Scheme VI). A weighted average I value (based on estimated proportions of tautomers) can be used; all atoms participating in the tautomerism must be calculated with a new I value:

For Scheme IV:
$$I(C) = 6.50, I(N) = 2.75$$
For Scheme V:
$$I(C) = 1.75, I(O) = 6.50$$
For Scheme VI:
$$I(N) = 2.75, I(O) = 6.50$$

$$R_2CH-CR=NR \quad \rightleftharpoons \quad R_2C=CR-NHR \qquad (IV)$$

$$
\begin{array}{ccc}
\mathrm{-CH_2-\overset{\displaystyle O}{\overset{\|}{C}}-} & \rightleftharpoons & \mathrm{-CH=\overset{\displaystyle OH}{\overset{|}{C}}-}
\end{array}
\qquad (V)
$$

$$
\begin{array}{ccc}
\mathrm{\overset{}{\underset{O}{\diagdown}}\!\!>\!\!-NH-} & \rightleftharpoons & \mathrm{\overset{}{\underset{HO}{\diagup}}\!\!>\!\!=N-}
\end{array}
\qquad (VI)
$$

VII. CONSIDERATION OF TYPES OF INDICES

Currently, there is a wide range of computable indices available for molecular structure representation. This presents opportunities for increased success in structure–activity investigations. Furthermore, the availability of sets of indices presents challenges for their optimum implementation. Some indices, such as the molecular connectivity chi indices, represent aspects of the whole molecule. Others, such as the kappa shape indices, encode elements of molecular shape. Even other indices, such as the polarity Q or the flexibility phi index, represent specific features of molecular structure.

Within the E-State formalism we have introduced four types of indices. Chapter 2 presented the derived formalism which leads to the E-State indices. Sometimes we have referred to these as atom-level indices. For a given molecular structure there is one atom-level index for each atom or hydride group. (The usual symbol for the E-State index of atom i is $s(i)$.) In addition, we have also introduced the E-State index for hydrogen atoms in a hydride group, such as —OH, —NH$_2$, and —CH$_3$. (The usual symbol for the hydrogen E-State index is $Hs(i)$.) In Chapter 4, we introduced the atom-type E-State indices. The atoms of a molecule are grouped according to a valence state classification scheme and the E-State values for each atom in a group are summed to give the atom-type E-State index for the atom type. There are also atom-type indices based on the hydrogen E-State as described in Chapter 4. In Chapter 10, we present an additional scheme for the creation of E-State index values for each bond type in the molecule.

In an investigation in which there is a common structure skeleton in every molecule in the series, it is profitable to consider the atom-level E-State indices along with all the other types of atom indices. This approach was described earlier in this chapter. When there is no such common structure element, then all the other types of indices are used as demonstrated in Chapter 8.

VIII. USE OF ORTHOGANALIZED INDICES: ANTAGONISM OF ADRENALIN BY 2-BROMO-2-PHENALKYLAMINES

A. INDEPENDENCE OF STRUCTURE DESCRIPTORS

A common way in which a QSAR model is obtained involves multiple linear regression. The property or activity is treated as the dependent variable and a set of descriptors are treated as the independent variables. In the theoretical basis for regression it is assumed that the descriptor variables are actually independent of each other; that is, orthogonal in the function–space sense. Such a set of orthogonal variables is seldom if ever available in practice for biological or chemical data. When such an orthogonal set is used, it is usually obtained by a mathematical procedure rather than obtained as an actual set of descriptors directly representing molecular structure (or property values).

We can state with some assurance that nature is not orthogonal. The features of nature which we can quantify and express as variables or descriptors are generally not unrelated to each other. This effect is particularly significant for a data set which consists of closely related molecular structures.

When intercorrelated variables are used in multiple linear regression (MLR) methods, certain problems can arise, including the following:

1. Values for the coefficients in the model equations are not stable with respect to addition or deletion of variables from the regression equation.

2. Values predicted for compounds not in the original data set may not be reliable, especially when the new compounds change the intercorrelation among the regression variables.

3. Errors estimated for regression coefficients on correlated variables are overestimated, sometimes significantly.

Several approaches to the nonorthogonality problem have been proposed to deal with this issue, including the following:

1. Convert the variables to a set of principal components in which each new variable is a linear combination of all the original variables [principal component analysis (PCA)]. The coefficients in the linear combinations are selected in total ignorance of the dependent variables. The first principal component is selected so as to maximize the percentage variance spanned by the variable. The remaining orthogonal variables are obtained with decreasing amounts of variance spanned. The criterion for the number of components selected is based on the percentage of variance covered, usually approximately 95–98%. It is the general experience that the principal component variables are not very

satisfactory QSAR correlates. PCA is a well-known procedure and will not be discussed.

2. Construct new orthogonal variables using the Schmidt orthogonalization procedure (Courant and Hilbert, 1931; Pilar, 1968; Randic, 1991). This approach will be illustrated in Section VIII,B.

3. The sum and difference of two variables may be considered as orthogonal under certain conditions. This method has been used by Hall (1990).

4. The method of partial least squares (PLS) also produces regression results using linear combinations which are orthogonal. A brief illustration of this method is provided in Section IX.

B. ANTAGONISM OF ADRENALIN BY 2-BROMO-2-PHENALKYLAMINES

In this section, we present the use of orthogonal variables obtained via the Schmidt orthogonalization process, as illustrated by a study of the antagonism of adrenalin by 2-bromo-2-phenethylamines (Scheme VII) (Graham and Kamar, 1963). The measure of biological response is the negative logarithm of the estimated dose for 50% response, pED_{50}. In the parent molecule (Scheme VII) Y and Z are hydrogen atoms; the four halogens and the methyl group comprise the substituents. The compounds with the observed, computed, and residual activities are given in Table 5.8.

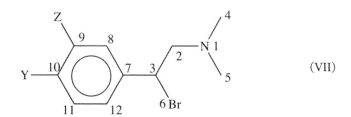

(VII)

Examination of the individual E-State values indicates that the indices for atoms 1–6 show very little variation throughout the data set. This effect arises because these atoms are quite remote from the region in which structure variation takes place. In terms of the E-State formalism, this remoteness is expressed as the number of atoms in the shortest path between a pair of atoms. In this case, the count of atoms is five or more. Hence, the perturbation is diminished by the factor $1/5^2$ (0.04) for the closest atom, No. 3. The other atoms exhibit an even greater diminution of effect.

Further examination indicates that the E-State indices for atoms 11 and 12 are significant in QSAR models of the activity (as will be shown later). However,

TABLE 5.8   Antagonism of Adrenalin by 2-Bromo-2-Phenethylamines

Observation	Substituent		pED$_{50}$	Calculated	Residual[a]	Pres[b]
1	H	H	7.46	7.93	−0.47	−0.70
2	F	H	8.16	8.16	−0.00	−0.00
3	Cl	H	8.68	8.78	−0.10	−0.11
4	Br	H	8.89	8.99	−0.10	−0.12
5	I	H	9.25	9.09	0.16	0.19
6	Me	H	9.30	9.11	0.19	0.23
7	H	F	7.52	7.72	−0.20	−0.32
8	H	Cl	8.16	8.04	0.12	0.15
9	H	Br	8.30	8.15	0.15	0.18
10	H	I	8.40	8.20	0.20	0.23
11	H	Me	8.46	8.21	0.25	0.29
12	F	Cl	8.19	8.32	−0.13	−0.18
13	F	Br	8.57	8.44	0.13	0.17
14	F	Me	8.82	8.50	0.32	0.39
15	Cl	Cl	8.89	8.94	−0.05	−0.05
16	Cl	Br	8.92	9.05	−0.13	−0.14
17	Cl	Me	8.96	9.11	−0.15	−0.17
18	Br	Cl	9.00	9.15	−0.15	−0.18
19	Br	Br	9.35	9.27	0.08	0.09
20	Br	Me	9.22	9.33	−0.11	−0.12
21	Me	Me	9.30	9.45	−0.15	−0.18
22	Me	Br	9.52	9.39	0.13	0.16

[a] Residual = pED$_{50}$ − calculated.
[b] Predicted residual based on the leave-one-out method.

the two E-State values are highly intercorrelated. Therefore, we decided to create a new variable which is the average of these two E-State indices: S11_12 = S(11) + S(12)/2. [Similar strategies have been used elsewhere (Vaughn and Hall, 1997).] This arrangement preserves the total information of s11 and s12 in the new combination variable.

The SAS system was used to determine the possible regression equations for this data set (SAS statistical software package, SAS Institute, P.O. Box 8000, Cary, NC 27511). We used SAS proc reg with the RSQUARE selection method to generate all possible one-, two-, and three-variable equations for QSAR models. We limited our attention to a maximum of three variables in order to keep the ratio of observations to variables at a reasonable level. Furthermore, principal component analysis of all the E-State indices present indicated that 95.2% of the variance is covered by three principal components. That is, there are at least three independent pieces of structure information in the E-State structure representation used here.

In the best one-, two-, and three-variable equations, there were three variables which were frequently found among the equations: s1112, s10, and s9. We decided to use these three variables as a basis for constructing three new orthogonal variables.

When the three variables are used in multiple linear regression analysis, the following equations are obtained:

$$pED_{50} = 6.37 \ (\pm1.19) + 1.17 \ (\pm0.50) \ S11\_12$$

$$r = 0.40, \ s = 0.53, \ F = 4, \ n = 22 \quad (5.1)$$

$$pED_{50} = 0.55 \ (\pm0.87) + 4.67 \ (\pm0.49) \ S11\_12 - 1.07 \ (\pm0.12) \ S(10)$$

$$r = 0.91, \ s = 0.25, \ F = 47, \ n = 22 \quad (5.2)$$

$$pED_{50} = -0.87 \ (\pm0.85) + 5.62 \ (\pm0.50) \ S11\_12$$

$$- 1.18 \ (\pm0.11) \ S(10) - 0.29 \ (\pm0.99) \ S(9)$$

$$r = 0.95, \ s = 0.20, \ F = 49, \ n = 22 \quad (5.3)$$

It is clear that the regression coefficients are unstable. With the addition of each new variable, the coefficients change significantly.

In the Schmidt orthogonalization process, one of the original variables is selected to remain unchanged. We selected the $S11\_12$ variable as the unchanged variable, giving it the symbol Ω_0. The second new variable, which is orthogonal to the first, is a linear combination of $S11\_12$ and $S(10)$ and is called Ω_1. The third orthogonal variable, called Ω_2, is a linear combination of $S11\_12$, $S(10)$, and $S(9)$ and is orthogonal to both Ω_0 and Ω_1.

The regression equations based on these three new orthogonal variables are as follows:

$$pED_{50} = 6.37 \ (\pm0.45) + 1.17 \ (\pm0.23) \ \Omega_0$$

$$r = 0.40, \ s = 0.53, \ F = 4, \ n = 22 \quad (5.4)$$

$$pED_{50} = 6.37 \ (\pm0.45) + 1.17 \ (\pm0.23) \ \Omega_0 - 1.07 \ (\pm0.10) \ \Omega_1$$

$$r = 0.91, \ s = 0.25, \ F = 47, \ n = 22 \quad (5.5)$$

$$pED_{50} = 6.37 \ (\pm0.45) + 1.17 \ (\pm0.23) \ \Omega_0$$

$$- 1.07 \ (\pm0.10) \ \Omega_1 - 0.29 \ (\pm0.09) \ \Omega_2$$

$$r = 0.95, \ s = 0.20, \ F = 49, \ n = 22 \quad (5.6)$$

Because the regression variables (Ω_0, Ω_1, and Ω_2) are independent, their coefficients do not depend on the presence or absence of other variables in the equation, and these coefficients are stable. The standard errors of the coefficients are also smaller than those in Eqs. (5.1)–(5.3).

This approach to QSAR, using a set of atom-level indices representing the same topological position in each of the molecules in the data set, is called topological superposition. Several detailed examples of this approach are given in Chapter 7. In the current case, the QSAR relation indicates that design attention can be focused on the phenyl ring atoms 9–11. The electrotopological character of these atoms is closely related to the biological activity, and this character is controlled by the substituents in the Y and Z positions. Manipulation of these substituents leads to modified activity which can be predicted from Eq. (5.6).

Table 5.8 lists the calculated, residual, and predicted residuals based on Eq. (5.6). These results are quite acceptable. Further they indicate the significance of the E-State indices as a representation of molecular structure and their ability to be used in this orthogonalization method.

IX. ILLUSTRATION OF THE PARTIAL LEAST SQUARES METHOD: FLAVONE DERIVATIVES AS HIV-1 INTEGRASE INHIBITORS

An alternative to developing orthogonal variables, as described in the previous section, is the use of the PLS method. This method was developed by Wold (1982) and has been implemented in software such as the SYBYL program (SYBYL software package, Tripos Associates, Inc., St. Louis, MO). In the PLS method orthogonal variables are developed as linear combinations of the original data variables. In contrast to the method of principal components, however, the coefficients in the new orthogonal variables are selected with full knowledge of the dependent variables. The new orthogonal variables, usually called latent variables, are selected one at a time until a minimum in the cross-validated error is obtained. Cross-validation is usually obtained using the leave-one-out method; each observation is deleted from the data set and predicted from the remaining $n - 1$ observations. In this manner, the dimensionally reduced orthogonal set of variables is constrained so as to minimize the communality of predictor and response variables.

A. FLAVONE DERIVATIVES AS HIV-1 INTEGRASE INHIBITORS

A set of flavone derivatives have been investigated by Engelman *et al.* (1991) as inhibitors of HIV-1 integrase and modeled by Boulamwini *et al.* (1995) using PLS. The structure of the flavones (Scheme VIII) reveals substituents at nine different positions, as also shown in Table 5.9, along with the measured activity for the pC values $[-\log (IC_{50})]$ for both cleavage and integration. Boulamwini

TABLE 5.9   Flavone Inhibitors of HIV-1 Integrase

Com-pound no.	Flavone	$-\log(IC_{50})$		Ring substituent								
		Cleav-age	Inte-gration	3	5	6	7	8	2'	3'	4'	5'
1	Quercetagetin	6.10	7.00		OH	OH	OH	OH			OH	OH
2	Baicalein	5.92	5.37		OH	OH	OH					
3	Robenetin	5.23	5.80	OH			OH			OH	OH	OH
4	Myricetin	5.12	5.60	OH	OH		OH			OH	OH	OH
5	Quercetin	4.63	4.87	OH	OH		OH			OH	OH	
6	Fisetin	4.55	5.07	OH			OH			OH	OH	
7	Luteolin	4.48	4.60		OH		OH			OH	OH	
8	Myricetrin	4.40	4.99	RH	OH		OH			OH	OH	OH
9	Quercetrin	4.22	4.41	RH	OH		OH			OH	OH	
10	Rhamnetin	4.21	4.54	OH	OH		OMe			OH	OH	
11	Avicularin	4.18	4.60	AR	OH		OH			OH	OH	
12	Gossypin	4.16	4.64	OH	OH		OH	GL		OH	OH	
13	Morin	4.12	4.50	OH	OH		OH		OH		OH	
14	6-Methoxyluteolin	4.03	4.41		OH	OMe	OH			OH	OH	
15	Kaempferol	4.01	4.19	OH	OH		OH				OH	

Note. Abbreviations used: RH, rhamanose; AR, arabinose; GL, glucose.

et al. report that examination of the data suggests that compound 14 is an outlier in both the cleavage and the integration data. Randomization control studies also support this conclusion. Raghaven et al. (1995) also found this compound to be an outlier when using a very different QSAR analysis, hence, compound 14 was eliminated from further analysis.

(VIII)

The E-State atom values were computed for the 17 common positions in the molecules and entered into PLS analysis. The number of PLS components found for each model (cleavage and integration) was determined by the number which minimized the cross-validated standard error. Three components were found for cleavage and five for integration.

B. PARTIAL LEAST SQUARES RESULTS

The following equations were obtained from the SYBYL software package for the two activities.

For cleavage:

$$pC = -0.149\ SO_1 + 0.017\ (SC_2 + 0.005\ SC_3) + 0.015\ SC_4$$
$$+ 0.250\ SC_5 - 0.888\ SC_6 - 0.123\ SC_7 + 0.138\ SC_8$$
$$+ 0.027\ SC_9 + 0.043\ SC_{10} + 0.014\ SC_{1'} + 0.107\ SC_{2'} \quad (5.7)$$
$$- 0.209\ SC_{3'} + 0.184\ SC_{4'} - 0.290\ SC_{5'} - 0.024\ SC_{6'}$$
$$- 0.421\ SO_4 + 11.475$$

$$r^2 = 0.975,\ s = 0.121,\ F = 131,\ n = 14,\ \text{c.v.}\ r^2 = 0.513$$

For integration:

$$pC = -0.189\ SO_1 - 0.023\ SC_2 - 0.140\ SC_3 - 0.015\ SC_4$$
$$+ 0.309\ SC_5 - 1.110\ SC_6 - 0.155\ SC_7 + 0.215\ SC_8$$
$$+ 0.016\ SC_9 + 0.043\ SC_{10} - 0.031\ SC_{1'} - 0.035\ SC_{2'} \quad (5.8)$$
$$- 0.380\ SC_{3'} - 0.048\ SC_{4'} - 0.236\ SC_{5'} - 0.052\ SC_{6'}$$
$$- 0.470\ SO_4 + 11.475$$

$$r^2 = 0.994,\ s = 0.093,\ F = 255,\ n = 14,\ \text{c.v.}\ r^2 = 0.734$$

Boulamwini et al. (1995) estimated the relative importance of each of the E-State indices in the models by examining the effect on the c.v. r^2 values when each of the variables was omitted one at a time. On this basis, it was concluded that the following ranking applies to the two activities:

For cleavage:

$$SC_6 \gg SC_{5'} > SO_4 > SC_5$$

For integration:

$$SC_6 \gg SC_{3'} > SO_4 = SC_5 > SC_3$$

These results are useful in the analysis of those compounds for drug design because they focus very specifically on atoms and/or regions of the molecules in the data set.

Although the catalytic core of HIV-1 integrase has been investigated by X-ray diffraction, no such studies have been done with an inhibitor cocrystallized with the enzyme. Hence, these atomic-level E-State indices can shed light on the nature of the relation between structure and activity. Polyhydroxylation appears to be required for activity (Fesen *et al.*, 1993; Burke *et al.*, 1995). In the current case, the most significant variables indicate that structure changes in the vicinity of the C_6, C_5, $C_{3'}$, $C_{5'}$, and O_4 positions are the most important in influencing activity. Changes that result in the decrease in E-State values in the vicinity of the C_6, $C_{3'}$, $C_{5'}$, and/or O_4 positions are predicted to enhance inhibitory power. These results indicate that placing an electronegative group such as an hydroxyl at position 6 will serve the purpose of increasing activity. Other substitution changes can serve to enhance activity to a somewhat lesser extent.

This study illustrates the ease with which such an investigation can be performed using PLS. It is clear that structure interpretation can lead to development of new drug molecules. A much more elaborate study done by Raghaven *et al.* (1995) used the comparative molecular field analysis three-dimensional method (Cramer *et al.*, 1988) but gave essentially the same conclusions as this more simply applied E-State method. It is clear that the PLS method can be applied to many types of data sets using the E-State indices. Another example will be given in Chapter 8.

X. ARTIFICIAL NEURAL NETWORKS: SOLUBILITY OF HIV REVERSE TRANSCRIPTASE INHIBITORS

The relationship between biological properties and molecular structure is often assumed to be linear, primarily because this assumption facilitates development of the relationship via multiple linear regression techniques. It is generally known, however, that the relation between structure and property is nonlinear. Since, in general, the exact nature of this relationship remains unknown, it is not possible to formulate a nonlinear regression model as a means for developing a QSAR relationship.

The development of artificial neural network (ANN) methodology along with accessible software has made ANN a potential pathway for obtaining nonlinear QSAR models (DeVillers, 1996; Egloff *et al.*, 1994; Hall and Story, 1996). We will not attempt to develop the fundamental principles of ANN here since there are several good references available and several papers have been published which describe the methodology (Zupan and Gasteiger, 1993; Egloff *et al.,* 1994; Hall and Story, 1996; Huuskonen *et al.,* 1997). We will illustrate one application of ANN using E-State indices.

A. SOLUBILITY OF INHIBITORS

The aqueous solubility of 25 reverse transcriptase inhibitors of HIV has been reported by Morelock *et al.* (1994). The parent skeleton structure (Scheme IX) is a three-ring structure with several heteroatoms and substituents in two locations. Positions *W, X,* and *Y* are occupied by the atoms C, N, and O in this data series. The structure together with the observed solubility values, as the logarithm, are given in Table 5.10.

TABLE 5.10   Solubility of Reverse Transcriptase HIV-1 Inhibitors

Obser-vation	Substituent W	X	Y	Z	R_1	R_2	log(S)	Calcu-lated[a]	Residual[b]
1	N	N(Et)	N	O	2-OMe, 2-OMe	H	5.153	5.021	0.132
2	N	N(tBut)	N	O	H	Me	4.849	4.958	−0.109
3	N	N(Et)	N	O	2-N(Me)₂, 4-Me	H	4.871	5.019	−0.148
4	C	N(Et)	N	S	H	Me	4.706	4.714	−0.008
5	N	N(cPr)	N	O	2,4-Me	H	4.272	4.244	0.028
6	C	O	N	O	2,4-Me, 8-NH₂	Me	3.762	3.759	0.003
7	C	O	C	O	H	Me	3.680	3.686	−0.006
8	C	N(Et)	N	O	H	Me	3.324	3.298	0.026
9*	C	O	N	O	2-Me, 8-NH₂	Me	3.043	3.592	−0.549
10	N	N(cPr)	N	O	H	Me	2.877	2.922	−0.045
11	N	N(Et)	N	O	H	Et	2.860	2.959	−0.099
12	N	N(Et)	N	O	H	Me	2.620	2.601	0.019
13*	N	N(Et)	N	S	H	Me	4.634	4.477	0.157
14	N	N(cPr)	N	O	4-Me	H	3.193	3.193	0.000
15	N	N(cPr)	N	CN[c]	4-Me		4.741	4.729	0.012
16	N	N(Et)	N	O	2-N(Me)EtOH	H	3.360	3.389	−0.129
17	C	N(Et)	O	O	H	Me	4.749	4.748	0.001
18	N	N(cBu)	N	O	H	Me	3.536	3.316	0.220
19	N	N(Et)	N	O	2,4-Me	H	4.554	5.014	−0.460
20	N	N(Et)	N	O	2-Cl	Me	4.114	4.103	0.011
21	N	N(Et)	N	O	2-Cl	H	5.360	4.991	0.369
22*	C	O	C	O	3-NH₂, 8-Me	H	3.926	3.853	0.073
23	N	N(Et)	N	O	2-Me, 4-CF₃	H	4.207	4.236	−0.029
24	CH	CH₂	CH	O	H	EtF	3.535	3.544	−0.009
25	CH	CH₂	CH	O	H	EtF₃	4.799	4.769	0.030

* Compounds used as the test set in the neural network training process.
[a] Solubility predicted from neural network.
[b] Residual = log(S) − calculated.
[c] —N(R₂)—C(=Z)— is replaced by —N=C(C≡N)—.

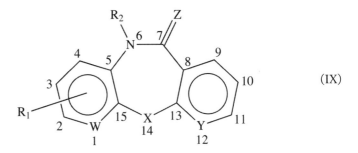

(IX)

Morelock *et al.* (1994) measured both the liquid chromatographic retention index and the melting point as parameters for estimation of the solubility. These two parameters were used in a MLR model of the solubility as follows:

$$\log(S) = -1.79 \ (\pm 0.15) \ \log k_s - 8.77 \ (\pm 1.00)$$

$$\times \ 10^{-3} \ \log \ (mp) - 0.59 \ (\pm 0.28)$$

$$r^2_{adj} = 0.88, \ s = 0.28, \ F = 85, \ MAE = 0.23, \ n = 25 \qquad (5.9)$$

This model is impressive in its standard error for measured water solubility. However, there is a serious difficulty with this type of relation. For prediction of drug molecules not yet synthesized, the two parameters are not available since they must be measured. Such a requirement is a serious limitation in drug design. For this reason, along with interpretability in terms of molecular structure, we turn to the E-State indices as a basis for modeling.

It is very useful to determine whether the E-State indices can be used to create a model for solubility, especially since the E-State indices are so readily computed and since they also lend themselves to structure interpretation. Since we expect that solubility may well be related to structure in a nonlinear way, we decided to subject these data to analysis using artificial neural networks. Preliminary examination by MLR methods suggested that certain atom-level E-State indices may be influential. Furthermore, certain atom-type E-State indices may also be influential. The list of indices is as follows: $S(2), S(6), S(10)$, $S(14), {}^4\chi^v_{PC}, SH^T(Csat), S^T(—NH_2), S^T(—NH—), S^T(—O—)$. The symbol $S(2)$ refers to the E-State index for the atom at position 2 in Scheme IX. The variable $SH^T(Csat)$ is defined as the sum of the hydrogen E-State indices for all $—CH_3$, $—CH_2—$, and $>CH—$ groups. The remaining atom-type E-State indices were defined in Chapter 4.

B. Neural Network Results

The architecture of the neural network adopted for this study is shown in Scheme X. The nine input nodes correspond to the nine structure descriptor variables listed previously. There are two hidden nodes and one output node, the measured aqueous solubility. The network is fully connected. The back-propagation algorithm is used as implemented in the Neuralyst Software from Cheshire Engineering Corporation (650 Sierra Madre Villa Avenue, Pasadena, CA 91107). The learning rate was set at 0.9 and the momentum at 0.7. The data set of 25 compounds was divided into a training set of 22 and a test set of 3, randomly selected by the software. Compounds 9, 13, and 22 are marked with an asterisk to indicate the test set in Table 5.10.

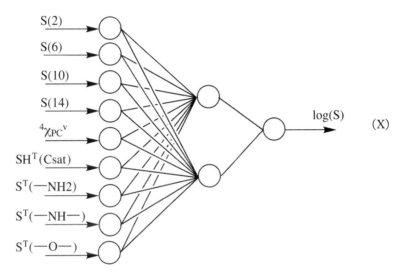

Several trial runs were done in order to select the learning parameters. The results of the best run are shown in Table 5.10. The statistics for this run are as follows: for the total set, $r^2 = 0.95$, MAE $= 0.11$, and rms $= 0.18$; for the test set, $r^2 = 0.88$, MAE $= 0.23$, and rms $= 0.30$. Although these results may not be optimized for the best network possible, it should be noted that the statistics for the test set are comparable to those obtained using measured physical parameters, as shown in Eq. (5.9). These results are considered excellent and indicate that the model obtained by ANN can be used for reliable prediction of aqueous solubility for compounds of this structure class.

In Chapter 9 other results from the ANN approach will be given. In each case it is clear that the E-State indices form a strong basis for molecular structure representation.

XI. SUMMARY

In this chapter, we discussed a range of topics dealing with strategies for approaching and executing structure–activity analyses. The E-State formalism has added greatly to the arsenal of weapons one may use in attacking such problems. The ability to implement atom-level indices permits a direct focus on issues related to pharmacophore identification. The combined use of atom-level, atom-type, and whole molecule indices provides a powerful approach to the wide range of problems confronting chemists who are designing molecules with specified properties. High-speed computation of E-State indices along with their direct interpretation in terms of structure make this structure representation a potent basis for molecule design.

XII. A LOOK AHEAD

In the following chapters, the strategies briefly described here will be illustrated in several ways. These illustrations will add to the meaning and insight of the strategies presented here. Chapter 6 deals with issues related to databases and virtual libraries—issues which make use of the concepts of molecular similarity and diversity.

REFERENCES

Boulamwini, J. K., Raghaven, K., Fresen, M., Pommier, Y., Kohn, K., and Weinstein, J. (1995). Application of the electrotopological state index to QSAR analysis of flavone derivatives as HIV-1 integrase inhibitors. *Pharm. Res.* **13**, 1892–1895.

Burke, Fesen, M., Mazumder, A., Wang, J., Carothers, A., Grumberger, D., Driscoll, J., Kohn, K., and Pommier, Y. (1995). Hydroxylated aromatic inhibitors of HIV-1 integrase. *J. Med. Chem.* **38**, 4174–4178.

Courant, R., and Hilbert, D. (1931). *Methoden der Mathematische Physik.* Springer, Berlin.

Cramer, R., Paterson, D., and Bunce, J. (1988). Comparative molecular field analysis (COMFA). 1. Effect of shape on binding of steroids to carrier proteins. *J. Am. Chem. Soc.* **110**, 5959–5967.

DeVillers, J. (Ed.) (1996). *Neural Networks in QSAR and Drug Design.* Academic Press, London.

Egloff, L. M., Wessel, M., and Jurs, P. (1994). Prediction of boiling points and critical temperatures of industrially important organic compounds from molecular structure. *J. Chem. Inf. Comput. Sci.* **34**, 947–956.

Engelman, A., Mizuuchi, M., and Craigie, R. (1991). HIV-1 DNA integration: Mechanism of viral DNA cleavage and strand transfer. *Cell* **67**, 1211–1222.

Fesen, M., Kohn, K., Leuteutre, F., and Pommier, Y. (1993). Inhibitors of human immunodeficiency virus integrase. *Proc. Natl. Acad. Sci. USA* **90**, 2399–2403.

Graham, J. D. P., and Kamar, M. A. (1963). Structure–action relations in N,N-dimethyl–2-halogen-ophenylamines. *J. Med. Chem.* **6**, 103–107.

Hall, L. H. (1990). Computational aspects of molecular connectivity and its role in structure activity modeling. In *Computational Chemical Graph Theory* (D. H. Rouvray, Ed.), pp. 216–217. Nova Science, New York.

Hall, L. H., and Story, C. T. (1996). Boiling point and critical temperature of a heterogeneous data set: QSAR with atom type electrotopological state indices using artificial neural networks. *J. Chem. Inf. Comput. Sci.* **36**, 1004–1014.

Huuskonen, J., Salo, M., and Taskinen, J. (1997). Neural networks for estimation of the aqueous solubility of structurally related drugs. *J. Pharm. Sci.* **86**, 450–454.

Morelock, M., Choi, L., Bell, G., and Wright, J. (1994). Estimation and correlation of drug water solubility with pharmacological parameters required for biological activity. *J. Pharm. Sci.* **83**, 948–952.

Pilar, F. L. (1968). *Elementary Quantum Theory*, pp. 63–65. McGraw-Hill, New York.

Raghaven, K., Boulamwini, J., Fresen, M., Pommier, Y., Kohn, K., and Weinstein, J. (1995). Three dimensional structure–activity relationship (QSAR) of HIV integrase inhibition: A comparative molecular field analysis (CoMFA). *J. Med. Chem.* **38**, 890–897.

Randic, M. (1991). Resolution of ambiguities in structure–property studies by use of orthogonal descriptors. *J. Chem. Inf. Comput. Sci.* **31**, 311–320.

Vaughn, T. A., and Hall, L. H. (1997). QSAR of phenol toxicity using electrotopological state and kappa shape indices. *Med. Chem. Res.* **7**, 407–416.

Wold, H. (1982). In *Systems Under Indirect Observation* (K. Joreskog, G. Wold, and H. Wold, Eds.), Part II, pp. 1–54. North-Holland, Amsterdam.

Zupan, J., and Gasteiger, J. (1993). *Neural Networks for Chemists*. VCH, Weinheim, Germany.

CHAPTER **6**

Database Applications: Molecular Similarity and Diversity

A commonly held view is that molecules which are structurally similar, by some criteria, may exhibit similar behavior with properties of comparable magnitude. In the context of biological properties, similarity presages comparable activities as ligands, substrates, inhibitors, or other agents engaged in intermolecular encounters in living systems. In the area of drug design, this is the guiding principal in structure–activity modeling. A mathematical relationship is sought between a measured activity and a molecular property or structure representation. The purpose is to predict another structure, representing a candidate molecule, and to shed light on the structure–activity relationship. The rationale is that similarity may portend comparable behavior. This principle is sometimes called the similar property principle (Johnson and Maggiora, 1990). The central element in this dialectic is similarity. This concept and associated methodology are explored in this chapter.

I. MOLECULAR SIMILARITY

In the words of T. S. Eliot, "all cases are unique and also similar." Molecules of different substances are indeed unique, but different molecules may elicit the same type of response, differing only in degree. It is tempting to impose our own intuitive or personal judgment on two objects and say that they are similar or dissimilar. These judgments are made on the basis of personal experience and are derived from our perceptions of form, function, or fluctuation (Testa and Kier, 1991). However, these are our perceptions, which may not coincide with the requirements of a biological receptor, enzyme, active site, or other effector responding to a molecule. To make progress in drug design or in any arena of structure–activity relationship, we must view a molecule not as a chemist but through the "eyes" of a fragment of a macromolecule surrounded by water in a living system. A means of expressing this kind of perception is hard to acquire, but such efforts are the goal of similarity considerations in any system in which function depends on molecular form and fluctuation.

There are a few criteria of similarity which we can state with a degree of confidence and still have some bases in chemical wisdom. Furthermore, it is not desirable to limit the realm of similarity by selecting a narrow basis for similarity such as a mechanistic basis. For example, it would be foolish to depart from concepts of structure developed around the presence and probable distribution of electrons. It is also wise to employ models in which the attraction between an electron and an atomic core influence the behavior of these ingredients. Other tenets of chemical wisdom are models of electron encounters, such as steric effects, conformational preferences, and intermolecular interactions. These are the major elements of the E-State, described in Chapter 2. The E-State concept is a prime candidate for the definition and description of similarity among molecules, molecular fragments, and atoms in molecules. One reason for this conjecture is that the E-State analysis produces numerical values which encode the extent of important structure attributes (described in Chapters 2 and 3). The structure information may be represented at the atom level (Chapter 2), the atom-type level (Chapter 4), and the bond-type level (Chapter 10) for both atom/hydride groups and separately for bonded hydrogen atoms, providing a broad basis for encoding molecular structure. The encoding of these attributes is a useful basis for similarity–dissimilarity quantitation, as we will develop in this chapter.

If molecular descriptors were to be derived as randomly assigned codes, as are street names in many cities, then a lookup procedure must be employed to find any of them in a large set of data (Fig. 6.1). There is no necessary connection between street name and location, and there is no rational basis for a locating algorithm. On the other hand, in some cities numbers are used for the east–west streets and letters are used for the north–south streets (Fig. 6.2)or a comparable pattern is used. In these places, it is possible to easily find the location of a street or, more specifically, an intersection. It is also possible to know the proximity of two locations directly from their codes. Translated into similarity terms, a large collection of objects (streets or molecules) must be named or encoded with some organization so that nearness may be established. Furthermore, selection and navigation can also be accomplished in a rapid and accurate manner. Naming and storing molecular information using codes not systematically related to structure produces a situation as shown in Fig. 6.1 in which browsing, locating, and establishing proximity are nearly impossible without a map (a time-consuming lookup table). Therefore, storing molecules in a database by alphabetical name or by physical property is of very limited value. We could use such a scheme only for lookup or for selection when only that physical property is of interest. This is a narrow view of similarity—one that would not likely produce the information needed in drug design, especially when operating with a typical industrial list of 500,000 or more entries. The E-State indices come into their own in this situation. Because of the way in

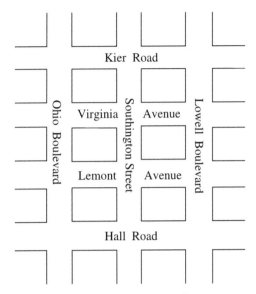

FIGURE 6.1   Street map of town with no rational scheme for street names and locations.

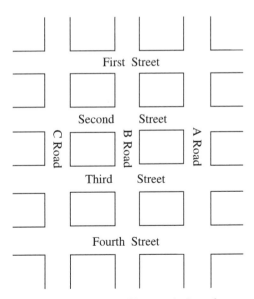

FIGURE 6.2   Street map of town with one possible rational scheme for names and locations.

which these indices are derived, there is a logical transition of information encoded from one structure to another, leading to an arrangement which may be analyzed and/or visualized. The browsing of a huge database for a particular structure based on E-State values may now be comparable to finding an intersection in Fig. 6.2.

II. THE E-STATE INFORMATION FIELD

A. THE INTRINSIC STATE

There is considerable information in the E-State indices, even in the intrinsic state values—information which can be used for the exploitation of a database of molecules. As an elementary example, we created a phase space of intrinsic state values, I, and the frequency of their occurrence in a molecule (Fig. 6.3). Three molecules are located in this space: ethyl acetate (Scheme I), alanine (Scheme II), and pyridine (Scheme III).

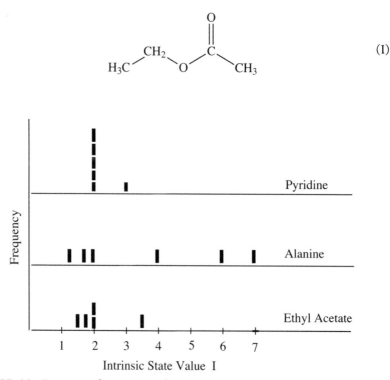

FIGURE 6.3 Frequency of occurrence of atom types versus intrinsic state value for three molecules.

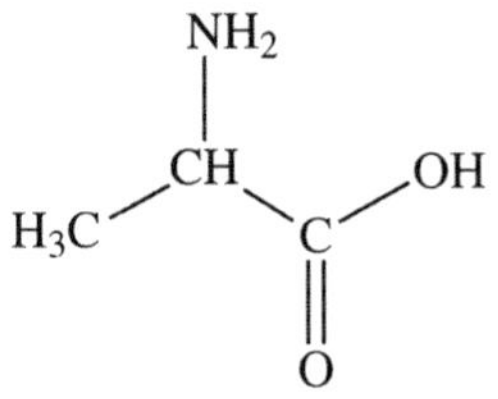

$$\underset{H_3C}{\overset{NH_2}{\underset{\underset{O}{\overset{|}{\|}}}{\overset{CH}{\diagdown}}}}\quad\overset{OH}{\underset{C}{\diagup}}$$

(II)

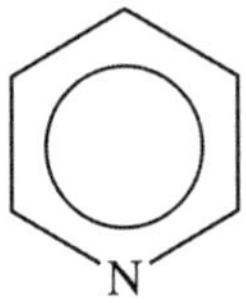

(III)

What information can be derived from this simple portrayal? We know where to look for a molecule such as pyridine with multiple atoms of the same hybrid state (five aromatic CHs). We also know that two of these molecules possess a common feature, a carbonyl group. There is an ether oxygen, an amine group, and an aromatic nitrogen atom, all in different molecules. We can browse or search through this simple molecular warehouse and find whatever is there. The essential feature of this phase space or data storage pattern is the coherence of the entries. At any position we know where we are and we know what atoms or groups are nearby or far away. If we move from left to right across Fig. 6.3, we move from buried or lipophilic features to mantle status atoms or groups with rich electronic features. This same message is portrayed in Fig. 3.2. The vertical dimension in Fig. 6.3 reflects the multiplicity of a feature; it is a region of intensity of any attribute exhibited by an atom or group.

This example illustrates the reality that the intrinsic state values and, consequently, the E-State values of a molecule constitute a profile—a numerical mosaic of information about the contents of a molecule and the interplay of their structure features. A set of E-State values is not a single molecular index encoding some aspects of structure. A database composed of molecules described by E-State indices is a manifold of surfaces connecting these values. The shape of these surfaces is unique (or nearly so) for each molecule. The surface shape or form encodes the attributes inherent in the intrinsic state and E-State as we described them in Chapter 3.

B. A Hyperspace of E-State Indices

The E-State values, summed or averaged for an atom type, may be thought of as numerical components of a space; that is, basis vectors in a manifold containing all possible molecular structures represented in terms of all possible atoms

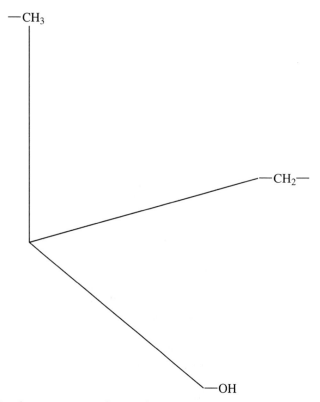

FIGURE 6.4   Three atom types used as coordinates in a structure space.

or groups. Each dimension is a parameter calculated for a particular atom or group. For example, E-State values for —CH_3, —CH_2—, and —OH constitute a three-parameter space as seen in Fig. 6.4. Consider a set of molecules composed exclusively of these groups (Tables 6.1 and 6.2). Two ways in which these structures can be displayed in a structure realm are described here.

1. Average Values of E-States for Atom Types

The first model uses the calculated average E-State value of each type of atom (or group) in a set of molecules, for example, those shown in Table 6.1. These values are displayed in Fig. 6.5 as an average value model. In this parameter space, the individual atom or group is located in a small segment of each axis in Fig. 6.5 based on the expected range of average values of E-States for each type of atom or group. The position may vary from group to group, but it can

TABLE 6.1 Average Atom-Type E-State Values for a Set of Alcohols, Glycols, and Alkanes

Molecule	$S^A(—CH_3)^a$	$S^A(—CH_2—)^b$	$S^A(—OH)^c$
$CH_3–OH$	1.00	0.00	7.00
$CH_3–CH_2–OH$	1.68	−0.25	7.57
$CH_3–CH_2–CH_2–OH$	1.93	0.60	7.88
$CH_3–CH_2–CH_2–CH_2–OH$	2.05	0.79	8.07
$HO–CH_2–CH_2–OH$	0.00	−0.13	7.62
$HO–CH_2–CH_2–CH_2–OH$	0.00	0.23	7.91
$CH_3–CH_2–CH_3$	2.13	1.25	0.00
$CH_3–CH_2–CH_2–CH_3$	2.18	1.32	0.00

[a] Symbol for the average of E-State values for —CH_3 groups in the molecule.
[b] Symbol for the average of E-State values for —CH_2— groups in the molecule.
[c] Symbol for the average of E-State values for —OH groups in the molecule.

usually be found in an expected region along each axis. Extending this model further, the average E-State values of atom types for each molecule will be scattered over a hyperspace in approximate regions that are predictable. From the E-State distributions in Table 6.3, it is apparent that the values for a specific atom or group can vary depending on its environment. They will, however, be positioned in a region within hyperspace—a region that is predictable. Any molecule is therefore represented by a constellation of vector components corresponding to the elemental contents of that molecule. Movement along any axis describes changing influences on the electrotopological state of an atom or

TABLE 6.2 Sum Atom-Type E-State Values for a Set of Alcohols, Glycols, and Alkanes

Molecule	$S^T(—CH_3)^a$	$S^T(—CH_2—)^b$	$S^T(—OH)^c$
$CH_3–OH$	1.00	0.00	7.00
$CH_3–CH_2–OH$	1.68	−0.25	7.57
$CH_3–CH_2–CH_2–OH$	1.93	1.19	7.88
$CH_3–CH_2–CH_2–CH_2–OH$	2.05	2.38	8.07
$HO–CH_2–CH_2–OH$	0.00	−0.25	15.25
$HO–CH_2–CH_2–CH_2–OH$	0.00	0.69	15.81
$CH_3–CH_2–CH_3$	4.25	1.25	0.00
$CH_3–CH_2–CH_2–CH_3$	4.36	2.64	0.00

[a] Symbol for the sum of E-State values for —CH_3 groups in the molecule.
[b] Symbol for the sum of E-State values for —CH_2— groups in the molecule.
[c] Symbol for the sum of E-State values for —OH groups in the molecule.

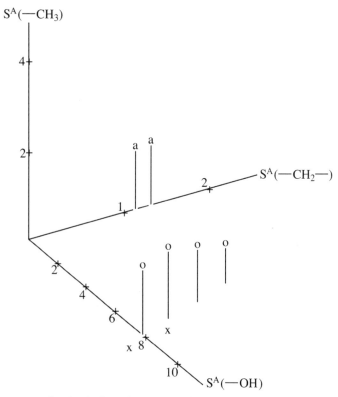

FIGURE 6.5   Several molecules located in a space defined by the average-value atom-type E-State indices. a, alkanes; o, alcohols; x, glycols.

group generated by proximal and distal electronegativity effects as well as the topological context of that atom or group.

In Fig. 6.5, a molecule is represented in a manifold of n dimensions, where n is the number of distinct types of atoms and groups. The position along the OH axis reflects varying influences operating on it within the molecule. Every molecule with an OH group would appear along the OH component direction at a nonzero location. Other features in that molecule would appear along other appropriate dimensions, with the position reflecting the environmental effects operative in the molecule. This pattern of distribution in hyperspace is navigable in a fashion similar to a warehouse in which products are stored near other products that are similar in the eyes of the warehouse manager. We can go to a location and find molecules that are similar to those that are nearby and different from those far away.

A measure of the proximity of molecules in Fig. 6.5 may be assessed in sev-

TABLE 6.4   Sum Atom-Type E-State Values for a Set of
Esters and Ketones

Molecule in Fig. 6.7	$S^T(=O)$	$S^T(>C=)$	$S^T(-O-)$
1	9.821	−0.211	4.403
2	10.678	−0.152	4.868
3	10.200	−0.104	4.501
4	10.804	0.077	4.938
5	10.123	0.182	4.632

group and one ether oxygen, sometimes associated in an ester group. The plot in Fig. 6.7 makes use of three atom types: the carbonyl oxygen, the carbonyl carbon, and the ether oxygen. For this plot the atom-type E-State indices are used in the summation form: $S^T(=O)$, $S^T(>C=)$, and $S^T(-O-)$. A major feature of this information, as revealed in Fig. 6.7, is the progression of the values for the carbonyl carbon, SdssC, from −0.2 to 0.2. The values are negative in the ester group because of the high intrinsic state value for both types of bonded oxygen atoms, leading to a very low value for the associated carbon atom. When the ether oxygen is separated from the carbonyl carbon atom, the SdssC value increases. This example clearly shows the arrangement of structures in the E-State space and the relation of this arrangement to electron accessibility, specifically at the carbonyl carbon. The two other variables also show significant variation with the structure changes in this set of structures.

III. CHARACTERIZATION OF E-STATE SPACE

In order to gain an understanding of the distribution of molecules in an E-State space, we created a miniature version of the full space composed of all atom-type E-State indices. For purposes of illustration, we selected 26 atom-type E-State indices and converted them to an orthogonal space using the principal components of the 26 variables (Hall and Kier, 1995; Walters *et al.*, 1998). The following are the 26 atom-type indices:
$S^T(-CH_3)$, $S^T(-CH_2-)$, $S^T(>CH-)$, $S^T(>C<)$, $S^T(=CH_2)$, $S^T(=CH-)$, $S^T(>C=)$, $S^T(...CH...)$, $S^T(...\overset{|}{C}...)$, $S^T(-NH_2)$, $S^T(-NH-)$, $S^T(>N-)$, $S^T(=N-)$, $S^T(...N...)$, $S^T(...NH...)$, $S^T(-\overset{\|}{N}=)$, $S^T(-OH)$, $S^T(-O-)$, $S^T(=O)$, $S^T(...O...)$, $S^T(-F)$, $S^T(-SH)$, $S^T(-S-)$, $S^T(...S...)$, $S^T(-Cl)$, and $S^T(-Br)$ (Hall and Kier, 1995).

The set of molecular structures was obtained by modifying and adding to the Pomona MedChem data set. We removed a small number of ionic and organometallic compounds, leaving 21,842 structures. We added the complete set of alkanes from ethane through dodecanes (663 structures), a set of aromatic hydrocarbons including alkylbenzenes, polyaromatic hydrocarbons (PAHs), and alkylated PAHs (515 structures), and a set of very highly branched alkanes (290 structures) for a total of 23,310 structures stored as SMILES strings (Hall and Kier, 1995). The atom-type E-State indices are very nearly linearly independent; $\geq 80\%$ of the pairwise correlation coefficients have $r < 0.05$ and only 1 has $r > 0.30$. Although these atom-type E-State indices are very nearly independent, the space is made exactly orthogonal by converting to the orthogonal space of the 26 principal components.

One way to characterize the distribution of structures in this orthogonalized space is to compute Euclidean-type distances among sets of related structures, according to an equation such as Eq. (6.1):

$$D = [\Sigma(x_{\text{ref}} - x_i)^2]^{1/2} \quad \text{sum over 26 principal} \atop \text{component coordinates} \quad (6.2)$$

A molecular structure of interest is called the reference structure. The Euclidean distances between the reference compound and the compounds in the data set are computed and the closest structures listed for examination. Our strategy here is to examine the space around common structure types to determine the similarity of nearest neighbors. First, we will examine simple hydrocarbons; then, we examine compounds with heteroatoms and some drug molecules.

A. HYDROCARBONS

The first set of simple molecules studied are alkanes, cycloalkanes, and alkyl-substituted benzenes. When a normal alkane such as octane is used as the reference structure, the nearest neighbors are heptane and nonane, both at a distance of 0.39 (arbitrary) units. Furthermore, no branched structures are found closer than 0.75 units. When we consider 3,3,4-trimethylhexane as the reference structure, the following six (trisubstituted) nonanes are found at a distance of approximately 0.04 units: 2,4,4-trimethylhexane, 2,3-dimethyl-3-ethylpentane, 2,2,3-trimethylhexane, 2,2-dimethyl-3-ethylpentane, 2,2,4-trimethylhexane, and 2,2,5-trimethylhexane. The next closest molecules are either octanes or decanes at distances of approximately 0.35 units. Likewise, when 3,3,4,4,-tetramethylhexane is the reference, only (tetrasubstitued) hexanes or pentanes are found close by, at distances ranging from 0.01 (2,2,3-trimethyl-3-ethylpentane) to 0.06 units (2,2,5,5-tetramethylhexane).

Similar results are found with monocyclic molecules. Using cyclohexane as the reference, cycloheptane and cyclopentane are the nearest neighbors at a distance of 0.40 units. For ethylcyclohexane, the nearest neighbors are butylcyclobutane at a distance of 0.04 units and methylcycloheptane at a distance of 0.05 units. One less carbon atom, as in methylcyclohexane, gives a distance of 0.40 units as does the addition of one carbon atom in ethylcycloheptane (0.40 units). When *ortho*-methylethylbenzene is the reference, the *meta* and *para* isomers are the closest neighbors at distances of 0.05 and 0.07 units, respectively. Molecules with one more carbon atom, such as 1,2-diethybenzene, are approximately 0.31 units away.

B. Molecules Containing Heteroatoms

Using 2-pentanol as a reference, 3-pentanol is at a distance of 0.08 units, 3-hexanol is at a distance of 0.26 units, and 2-butanol at a distance of 0.31 units. The nearest neighbors to 1,2-dichloroethane are dichloromethane at a distance of 0.27 units, 1,3-dichloropropane at a distance of 0.32 units, and 1,2-dichloropropane at a distance of 0.59 units. If the similarity distance were based only on $S^T(—Cl)$, $S^T(—CH_2—)$, and $S^T(>CH—)$, then 1,2-dichloroprpane would be much closer to 1,2-dichloroehtane.

When 4-ethylphenol is the reference, the 3- and 2-isomers lie at a distance of 0.05 and 0.13 units, respectively. Methyl- and propyl-substituted compounds lie at distances of approximately 0.30 units. Similarly, for *m*-chlorophenol; the *para* isomer lies at a distance of 0.08 units and *ortho* at 0.15 units; *m*-chlorobenzylalcohol is at a distance of 0.49 units.

C. Generalizations

On the basis of many observations such as those described noted previously, it is possible to present a set of generalizations on the nearest neighbors in the E-State space. These statements strongly suggest that molecules with structures perceived to be similar are located close to each other:

1. Nearest neighbors tend to have the same or very similar skeletal structure. For alkanes, cyclic and acyclic, the nearest neighbors to an unbranched reference are also unbranched. Branched alkanes are the nearest neighbors to a branched alkane and the nearest neighbors have the same branching pattern.

2. Nearest neighbors tend to have the same set of heteroatoms.

3. Nearest neighbors tend to have the same number of atoms.

4. There is a pattern in the computed similarity distances. Although the magnitude of the distances depends on the dimensionality of the E-State space, the following data illustrate the concepts. For the 26-dimension space described previously, molecules which are perceived to be very similar are very close, generally up to 0.10 units. Molecules which are similar but which possess

one more (or less) carbon atom tend to lie at somewhat larger distances, usually exceeding 0.25–0.30 units. Isoskeletal molecules with differences in heteroatoms lie at even greater distances.

D. DRUG MOLECULES

We have also examined the similarity characteristics of four drug molecules. We computed the similarity distance for each of the four in the 23,310 compound data set. We do not expect to find as many very similar structures as found for hydrocarbons or the other simple molecules discussed previously. The set of structures in the data set is much too limited.

The results for the four drug molecules are shown in Tables 6.5–6.8. The first example for searching a database is the drug molecule mefloquine. The first candidate molecule shown in Table 6.5 is similar to the reference, mefloquine; three other candidates are also given. In all these examples, the distance computation is based on all 26 atom-type E-State indices. That is, all atom types are given equal weight. It is not necessary to use all the atom types in the whole space or, indeed, all the atom types in the reference molecule. In fact, using atom types not present in the reference tends to exaggerate the distances. Furthermore, it may well be that some atom types are not important for the particular biological activity (or physical property) under consideration. For example, in this case it may well be that the —CH$_2$ and —CH$_3$ groups are not very important for antimalarial activity. If these two groups are eliminated, then the last molecule in Table 6.5 is computed to be at a much smaller distance. In this fashion, the user can fine-tune the similarity measure.

Another example is meperidine, for which several similar candidates were found and are shown in Table 6.6. The similarity is apparent to the eye in this case. In Table 6.7, five candidates are listed when niridazole is the reference molecule; the first candidate is similar to niridazole. There were no very similar molecules in the data set, so only one was found at a close distance. Finally, minoxidil in Table 6.8 is used as a reference molecule. One molecule is found to be somewhat similar; three other candidates are also listed. All the molecules have the same nitrogen heteroatomic ring system. The character of the parameter space in which the reference molecular structure is embedded may be altered based on:

1. The information available on the reference molecule
2. The objectives of the search
3. The caveats drawn from the molecules relative to pharmacokinetcs, metabolism, and toxicity problems (Walters *et al.*, 1998; van Drie and Lajiness, 1998)

TABLE 6.5 Database Search with the Reference Compound Mefloquin

Molecular structure	Distance (arbitrary units)
	0.23
	0.74
	0.75
	1.13

TABLE 6.6 Database Search with the Reference Compound Meperidine

Molecular structure	Distance (arbitrary units)
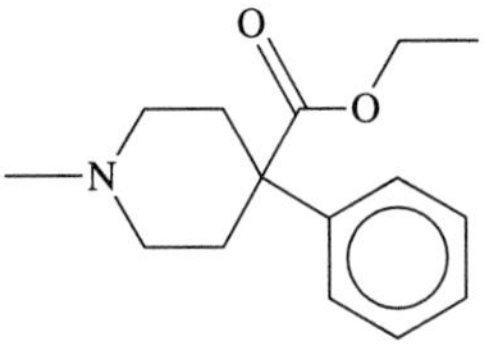	
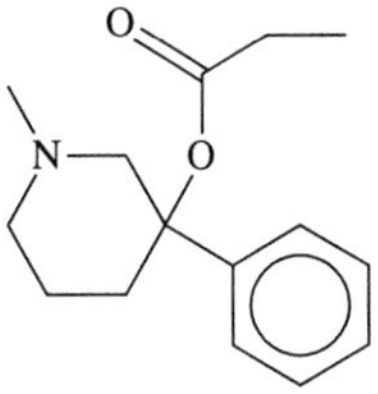	0.12
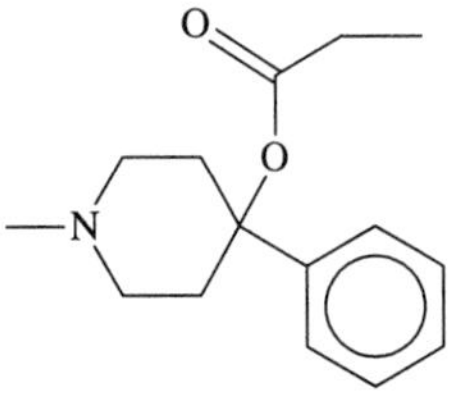	0.14
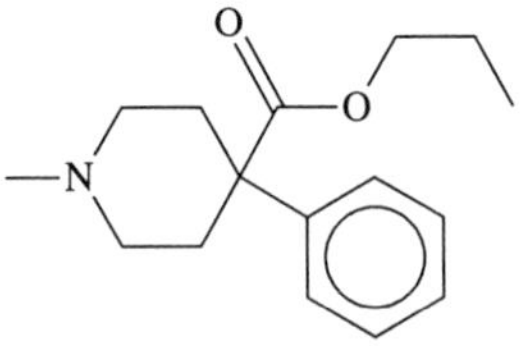	0.26
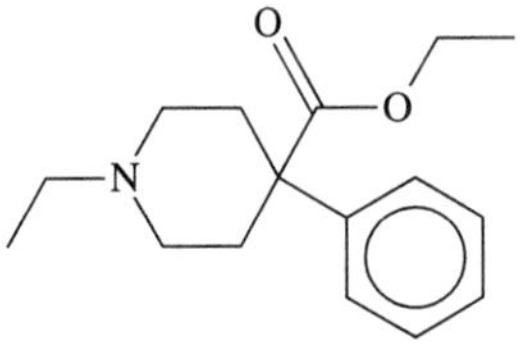	0.31
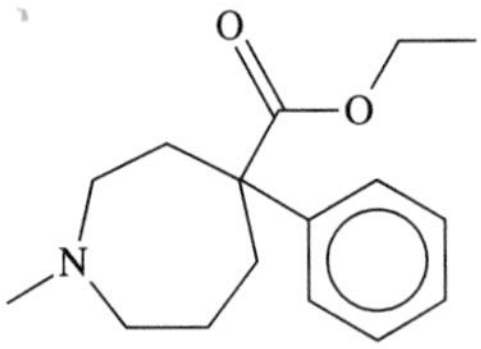	0.31
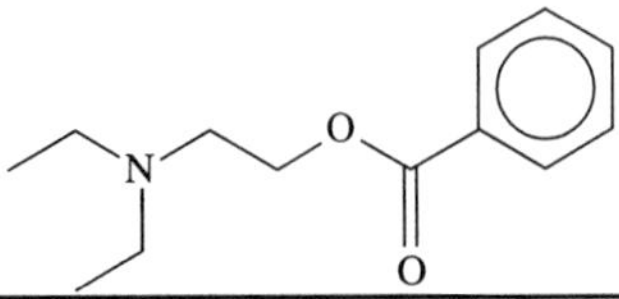	0.53

TABLE 6.7 Database Search with the Reference Compound Niridazole

Molecular structure	Distance (arbitrary units)
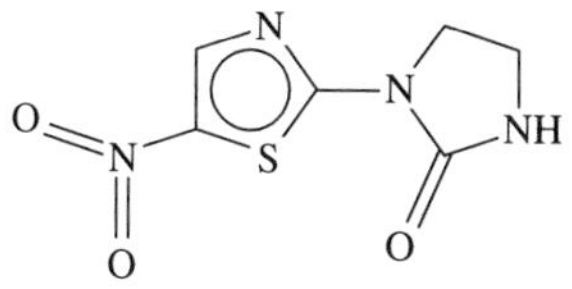	
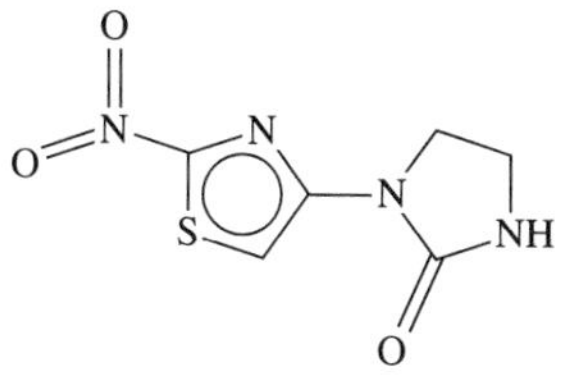	0.25
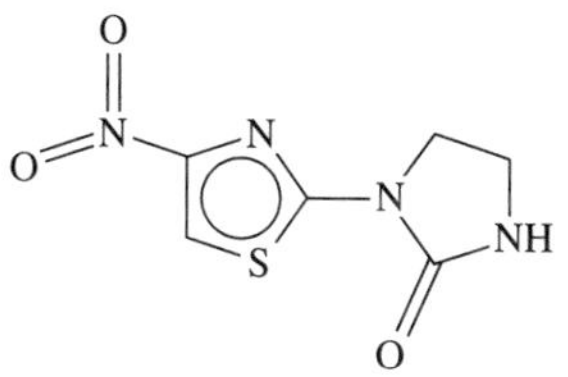	0.85
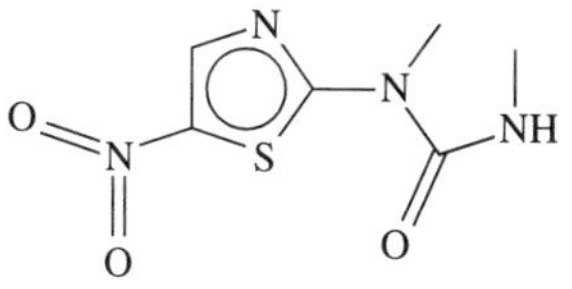	0.99
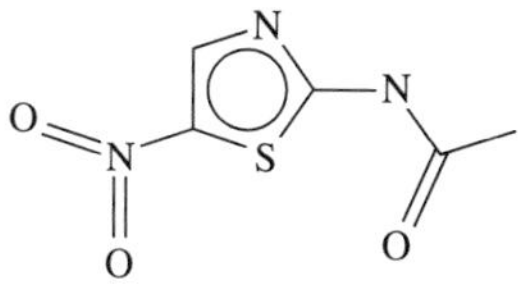	1.20
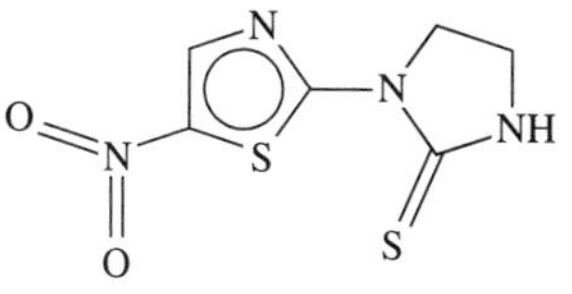	1.50

TABLE 6.8 Database Search with the Reference Compound Minoxidil

Molecular structure	Distance (arbitrary units)
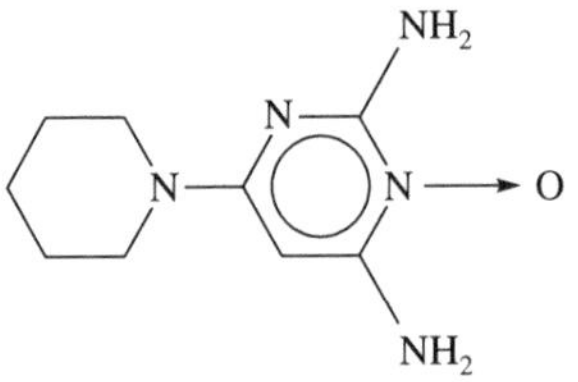	
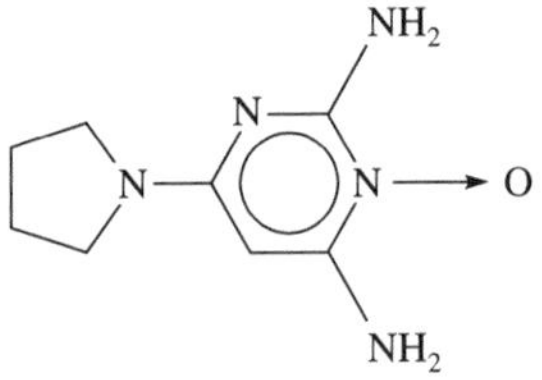	0.33
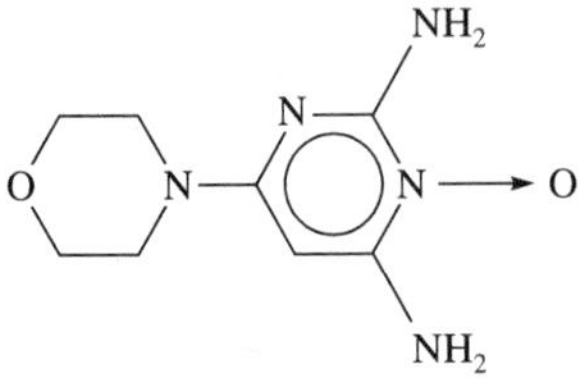	1.23
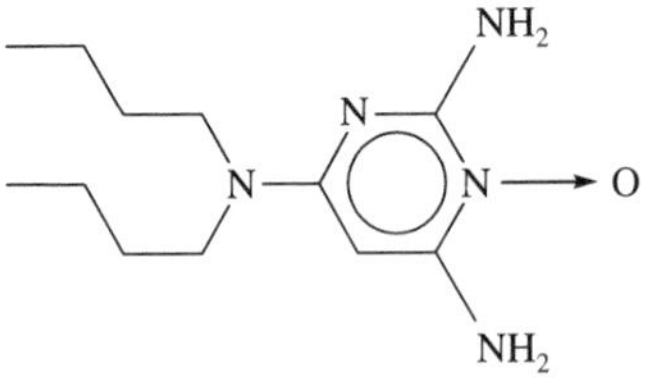	1.31
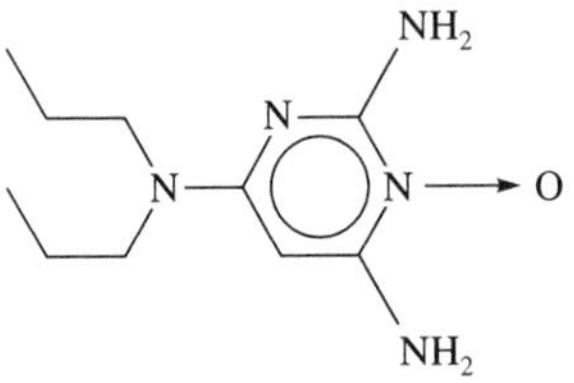	1.33

IV. FUNCTIONAL GROUP DESCRIPTION IN PARAMETER SPACE

Either the addition model or the sum model may be employed in the character-ization or the search for functional groups or variants of them in a database. For example, the ester group, $-O-(C=O)-$, or its variants or ingredients may be encoded by the sum model and positioned in a parameter space for recall or for searching. An example of the sum model of the ester group is shown in Fig. 6.7. Five molecules bearing the three ingredients of the ester group are located in three-dimensional space bounded by the average E-State values for the atoms $-O-$, $=O$, and $>C=$. The molecules are an aliphatic ester, an aromatic lactone, an α,β unsaturated lactone, and structures with a separated pair of functions, such as methoxy and aldehyde groups, and a methoxy group together with a ketone.

The locations of these molecules is dependent on the average values of the three E-State indices. It is possible to design a search for any one of these struc-tural features. Such a search is designed from information generated by cal-culations of typical molecules of each type and then scanning that part of pa-rameter space with the selected values. The E-State indices are generated for a specific molecule, such as isopropyl benzoate. The E-State index values for the three atoms of the ester group are used for the computation of distances. In this fashion, the closest molecules will all have an ester function and also have elec-tron accessibility characteristics like those of the reference, isopropyl benzoate. If it is desired that the molecules found also have a $>CH-$ group attached to the $-O-$ of the ester group, then the atom-type E-State index for the $>CH-$ group can be added to the search variables. In this fashion, the user may design the search characteristics in a straightforward chemical manner. The "search parameters" are simply the E-State index values for those atoms of interest in the molecule of interest.

V. THE QUANTITATION OF MOLECULAR FRAGMENTS

In the design of new and improved drugs, we adopt the principle that molecules with similar structural characteristics may have similar biological responses. This principle of similarity may be defined in terms of atom or group attributes and their spatial interrelation. We say that a molecule with a constellation of structural features complimentary to that of a defined pharmacophore should have a comparable biological activity. This principle is one basis for the design and synthesis of, or the database searching for, candidate molecules in the drug design process. One area of interest is the ability to define precisely molecular patterns embedded in the pharmacophores. A coding of these fragments provides a measure of similarity between molecules and also becomes a reference for searching. This section addresses the issue and proposes a simple and automated way of encoding fragment and pharmacophore structures.

A. Method

A fragment in a molecule is defined in its simplest form as a substructure in which two atoms or groups are separated by a certain number of chemical graph atoms comprising a path. The atoms or groups are described by their atom type and valence state. The connecting path is defined as the smallest number of contiguous chemical graph atoms connecting and including the terminal atom. The fragment in Scheme IV illustrates this point.

(IV)

We designate this fragment using the terminal atoms as codes in addition to a numerical value for the count of atoms in the path. We specify this particular fragment in Scheme IV as $5[(-NH-):(=O)]$. The ingredients or variables of the fragment designation are therefore:

1. A secondary N sp^3 amine group, $-NH-$
2. An O sp^2 carbonyl oxygen atom, $=O$
3. A count of five atoms in the path separating (and including) the terminal atoms in the chemical graph fragment

One approach to the derivation of a numerical code specifying this fragment is to utilize the E-State paradigm. From this method, we can calculate the contribution of one atom to a reference atom within the molecular fragment based on its perturbation of the intrinsic state. In the case of the fragment shown in Scheme IV, we use the calculation of the E-State values of the terminal atoms of the fragment, $-NH-$ and $=O$. We calculate the specific perturbation of one terminal atom in this fragment on the other terminal atom. Thus, for the calculation of the influence of the $-NH-$ on the $=O$, we use the algorithm

$$F = \frac{I(=O) - I(-NH-)}{r^2}$$

The symbol F is adopted for this particular application as the numerical value of the fragment. Using $I(-NH-) = 2.5$, $I(O) = 7.0$, and $r = 5$, we calculate for the perturbation of the $-NH-$ on the $=O$ atom: $F = 0.1800$. It is apparent that the perturbation of the $=O$ atom on the $-NH-$ group is -0.1800. The positive value is adopted in all cases as the characteristic F_{ij} value for any fragment. This calculation simultaneously produces a code for the entire fragment since the three variables of the fragment listed previously are employed in the algorithm. The number 0.1800 is thus a specific number encoding the fragment $5[(-NH-):(=O)]$. This number is one of the set of numbers calculated and used to sum the total perturbation associated with the E-State value of each atom. As a result, the F index value 0.1800 can be searched for and recalled whenever the fragment for which it is specific is found in any molecule.

The following question arises: Is the F value 0.1800 unique for the fragment shown in Scheme IV? Without further analysis it cannot be positively stated; however, our intuition based on experience tells us that molecules in a database containing some specific fragment or constellation of atoms can now be quickly identified using existing, in-place technology. A few examples illustrate this method.

1. Group Specification

A functional group can be identified in any molecule merely by searching for the appropriate F value or set of F values which correspond to it. An example is the carboxyl group shown in Scheme V.

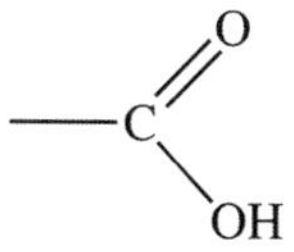

$$-C{\overset{\displaystyle \nearrow \text{O}}{\underset{\displaystyle \searrow \text{OH}}{}}} \qquad\qquad (V)$$

There are three fragments in the group coded using the convention described previously:

F_1 $3[(=O):(-OH)]$

F_2 $2[(=O):(>C=)]$

F_3 $2[(-OH):(>C=)]$

The computed values of F are as follows:

$$F_1 = \frac{I(=O) - I(OH)}{(3)^2} = \frac{7 - 6}{9} = 0.111$$

$$F_2 = \frac{I(=O) - I(=C<)}{(2)^2} = \frac{7 - 1.67}{4} = 1.333$$

$$F_3 = \frac{I(-OH) - I(=C<)}{(2)^2} = \frac{6 - 1.67}{4} = 1.083$$

These three F values, representing the ΔI values for the three fragments in a carboxyl group, form a constellation of codes which characterize the whole functional group. Furthermore, this functional group may be searched among a database of molecules using these three values. These values uniquely define the carboxyl functional group. All these groups in a database will be extracted and no molecule without this group will be tagged.

A second example lays the groundwork for group variation. Consider the phenyl group shown in Scheme VI.

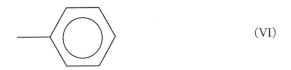

(VI)

There is one way that this can be defined in terms of F values—using the ipso or attached atom as the reference point:

$$F_1 \text{ (for the 1,2 and the 1,6 fragments)} = \frac{1.666 - 2.000}{4} = -0.1111$$

$$F_2 \text{ (for the 1,3 and the 1,5 fragments)} = \frac{1.666 - 2.000}{9} = -0.0370$$

$$F_3 \text{ (for the 1,4 fragment)} = \frac{1.666 - 2.000}{16} = -0.0208$$

For convenience, the positive values are used. In the case of the phenyl group, there will be a set of two F_1, two F_2, and one F_3 fragments in any molecule containing an unsubstituted phenyl group. This constellation is unique for this group.

TABLE 6.9   *F* Values of Monosubstituted Phenyl Groups

Isomer graph	Fragment	No.	*F* Value
ortho	1,2	2	0.0833
	1,3	4	0.0370
	1,4	2	0.0208
meta	1,2	4	0.0833
	1,3	2	0.0370
	1,4	2	0.0208
para	1,2	4	0.0833
	1,3	4	0.0370
	1,4	0	0.0000

2. Position Isomers

Phenyl substituents as parts of molecules in a database may be more complicated than the simple case encoded previously. A phenyl substituent may be mono-substituted in three different ways. It is possible to describe and hence search for any of these isomers using the *F* values as guides. Consider a set of phenyl derivatives that are *ortho*, *meta*, or *para* substituted as in Table 6.9. These sets of values clearly indicate the differentiation among the three isomeric possibilities.

3. Homologs

Finding members of a homologous series may well be an objective in a database search. The identification of a homologous series is relatively easy using the method previously described. Consider the reference fragment $>N—C—C—C—OH$. The calculations of the F_{N-OH} value are

$$F_{N-OH} = \frac{I(OH) - I(N)}{r^2} = \frac{7 - 2}{25} = 0.2000$$

The next higher homolog, $>N—C—C—C—C—OH$, is encoded by

$$F_{N-OH} = \frac{7 - 2}{36} = 0.1389$$

In contrast, the *nor* modification (one lower homolog) is described by

$$F_{N-OH} = \frac{7 - 2}{16} = 0.3125$$

In general, a homologous series varying in one fragment, i and j, is represented by the equation

$$F_{ij} = \frac{I_i - I_j}{r^2}$$

where r is the count of atoms separating and including i and j.

4. Pharmacophore Patterns

One of the most active areas of database management is in the search for molecules with structural features mimicking a pattern in a reference molecule considered to be a pharmacophore. The method described here is very useful for this objective. For example, consider the pharmacophore originally proposed for histamine as in Scheme VII.

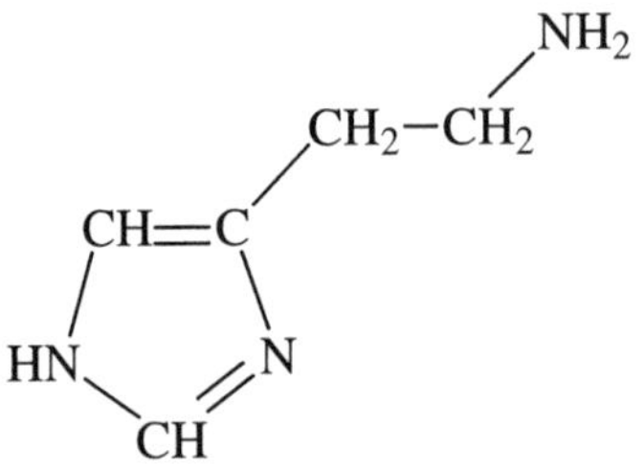

(VII)

The three nitrogen atoms, each in a different hybrid and/or hydride state, form a pattern of three fragments:

1. NH_2—C—C—C—C—NH
2. N—C—NH
3. NH_2—C—C—C—N

The coding for this pharmacophore embraces the three fragments, calculated to be

$$F_1 = \frac{I(NH_2) - I(NH)}{r^2} = \frac{4.00 - 2.50}{36} = 0.04167$$

$$F_2 = \frac{I(N) - I(NH)}{r^2} = \frac{3.00 - 2.50}{9} = 0.05556$$

$$F_3 = \frac{I(NH_2) - I(N)}{r^2} = \frac{4.00 - 3.00}{25} = 0.04000$$

Note that the ordering of the terms in each fragment is designed to give positive values for the F index. This is an arbitrary choice since there are negative and positive values for each fragment value in all molecules. The search for the histamine pharmacophore is now possible using this constellation of F values in a database search.

B. VALUE OF THE FRAGMENT SEARCH APPROACH

It is possible with the procedure described previously to calculate a numerical code (F value) characteristic of a fragment defined by two atoms or groups separated by any number of atoms in a chemical graph. Such an index may be unique for any fragment; this is our current conjecture. Nevertheless, such an index makes it possible to search through E-State calculated values for any molecule and to identify those containing any designated fragment. The search using this procedure can be expanded to encode the value of a pharmacophore or other constellation of atoms or groups. It is thus possible to conduct a rapid search through a database to find a pharmacophore imbedded in any molecule. The method uses existing E-State calculation algorithms currently in place and operational.

VI. ORGANIZATION OF DATABASE SUBSETS

Within database subsets of molecules there are patterns of structure variation that are of interest in compound design. These patterns characterize the relative similarity or diversity within the subset. The ability to organize the subset in some manner facilitates the design of other modifications, selection of diverse structures for testing, cluster analysis based on structure, or structure–activity analyses. To illustrate this organization and how the E-State indices can accomplish this, we examine the notorious polychlorobiphenyls (PCBs; Scheme VIII).

(VIII)

Three atom types are present in this series, each designated by its atom-type E-State code. These atom types are the chlorine atom (—Cl), the aromatic —CH— group (...CH...), and the substituted aromatic carbon atom —C< (...C...). (The reader is referred to Chapter 4 for descriptions of the atom-type indices and their identifying codes.) Using the sum of E-States for each atom type, a descriptor space is created. To illustrate the structure organization possible, the two parameters, $S^T(-Cl)$ and $S^T(...CH...)$, are used to create a two-dimensional space or a projection from a higher dimension space. A view of this space over a relatively wide range reveals many of the possible PCBs from the unsubstituted through the tetrasubstituted derivatives (Fig. 6.8). The major parameter governing the position in this space is the number of chlorine atoms. To extract more useful information, we examine subsets with common numbers of chlorine atoms by viewing a restricted range of parameter values.

In Fig. 6.9 the dichlorobiphenyls plotted in the atom-type E-State space (sum of E-State values) for Cl and aromatic —CH—, coded as $S^T(-Cl)$ and $S^T(...CH...)$, are shown. The structures with their letter codes are shown in Table 6.10. Figure 6.9 presents an opportunity to assess the structure information as organized by these two parameters. The disubstituted biphenyls are arrayed in a pattern that can be interpreted in structure terms. The upper ellipse

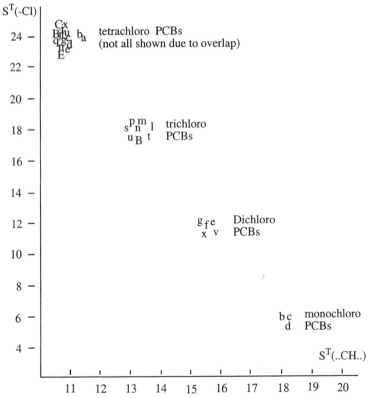

FIGURE 6.8 A plot of PCB compounds in a space defined by two atom-type E-State indices. Shown here are monochloro- through trichloro-PCBs.

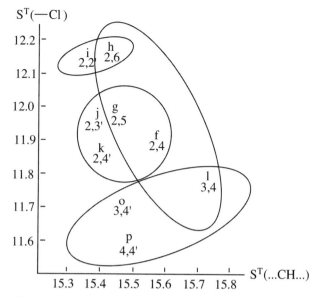

FIGURE 6.9 Plot of dichlorobiphenyls in a space defined by two atom-type E-State indices. The letter codes are defined in the text. The ellipses of enclosure are also defined in the text.

contains molecules h and i, which are substituted on the rings at the 2 or $2'$ positions in Scheme VIII. The lowest ellipse contains molecules with chlorine atoms far removed from the juncture of the two rings. In between these two subsets lie encircled biphenyls substituted at intermediate positions. The largest ellipse encloses PCBs with substituents on only one ring. This set of PCBs is clearly organized in a meaningful way in terms of molecular structure.

The organization of these molecules in this parameter space is more evident in the case of trichlorobiphenyls shown in Fig. 6.10. An examination of this figure shows a cluster of derivatives with all three chlorine atoms on the same ring. These are labeled t, s, r, and q corresponding to the derivatives 2,3,6, 2,4,6, 2,3,5, and 2,3,4. Within this cluster, the molecules with chlorine atoms closer to the ring juncture in Scheme VIII are located higher in the plot, ranking in the previously given order. As can be seen, the pattern of organization for trichloro compounds is similar to that for the dichloro compounds.

A second pattern in the grid of trisubstituted PCBs is derived from an examination of the location of compounds substituted in the positions near the ring junctures. This includes the 2,3,6 and the 2,4,6 derivatives previously discussed and also the 2,3,2' and 2,4,2' derivatives. On the other hand, if substituted positions are far from the ring junctures, as in the 3,4,4', 3,5,4', and 3,4,3' derivatives, these molecules are located in the lower region of the grid. Mole-

TABLE 6.10   E-State Indices for Polychlorobiphenyls

Obser-vation	Symbol	PCB	Chlorine count	$S^T(—Cl)$	$S^T(...CH...)$	$S^T(...\overset{|}{C}...)$
1	a	Biphenyl	0	0.000	20.781	0.0000
2	b	2-Cl	1	6.059	18.005	0.7997
3	c	3-Cl	1	5.895	18.078	0.7790
4	d	4-Cl	1	5.799	18.114	0.7775
5	e	2,3-diCl	2	12.028	15.604	1.2022
6	f	2,4-diCl	2	11.900	15.527	1.3457
7	g	2,5-diCl	2	11.980	15.429	1.4249
8	h	2,6-diCl	2	12.159	15.418	1.3680
9	i	2,2'-diCl	2	12.144	15.355	1.4456
10	j	2,3'-diCl	2	11.972	15.392	1.4696
11	k	2,4'-diCl	2	11.871	15.405	1.4959
12	l	3,4-diCl	2	11.768	15.713	1.1800
13	m	3,5-diCl	2	11.831	15.565	1.3266
14	n	3,3'-diCl	2	11.803	15.442	1.4768
15	o	3,4'-diCl	2	11.704	15.463	1.4937
16	p	4,4'-diCl	2	11.607	15.488	1.5050
17	q	2,3,4-triCl	3	17.942	13.429	1.3717
18	r	2,3,5-triCl	3	17.991	13.217	1.5959
19	s	2,3,6-triCl	3	18.155	13.144	1.6166
20	t	2,4,6-triCl	3	18.042	13.130	1.6825
21	u	2,3,2'-triCl	3	18.131	13.045	1.7389
22	v	2,3,3'-triCl	3	17.954	13.059	1.7908
23	w	2,3,4'-triCl	3	17.851	13.057	1.8357
24	x	2,4,2'-triCl	3	17.999	12.945	1.9103
25	y	2,4,3'-triCl	3	17.823	12.967	1.9528
26	z	2,4,4'-triCl	3	17.721	12.969	1.9920
27	A	3,4,2'-triCl	3	17.859	13.095	1.7893
28	B	3,4,3'-triCl	3	17.687	13.130	1.8150
29	C	3,4,4'-triCl	3	17.586	13.140	1.8447
30	E	3,5,2'-triCl	3	17.927	12.969	1.9080
31	F	3,5,3'-triCl	3	17.753	12.996	1.9431
32	G	3,5,4'-triCl	3	17.651	13.002	1.9785

cules forming a central cluster in the grid are the derivatives of both rings, including a 2,3 substitution pattern. Below these are the mixed ring derivatives with 3,3' substituents always present.

The tetrasubstituted derivatives of PCB are shown in Fig. 6.11. The same general organization of substitution patterns is observed here. In addition, the derivatives that have two chlorines on each ring are found in a segment of the plot on the left margin. As expected, the 2,6,2',6' derivative is near the top of

TABLE 6.10 (*continued*)

Observation	Symbol[a]	PCB	Chlorine count	$S^T(-Cl)$	$S^T(...CH...)$	$S^T(...\overset{\vert}{C}...)$
33	a	2,3,4,5-tetraCl	4	23.980	11.345	1.3888
34	b	2,3,4,6-tetraCl	4	24.111	11.159	1.5545
35	c	2,3,4,2'-tetraCl	4	24.060	10.938	1.8271
36	d	2,3,4,3'-tetraCl	4	23.880	10.936	1.8975
37	e	2,3,4,4'-tetraCl	4	23.774	10.923	1.9552
38	f	2,3,5,2'-tetraCl	4	24.113	10.749	2.0234
39	g	2,3,5,3'-tetraCl	4	23.931	10.740	2.1033
40	h	2,3,5,4'-tetraCl	4	23.824	10.723	2.1665
41	i	2,3,6,2'-tetraCl	4	24.285	10.713	1.9994
42	j	2,3,6,3'-tetraCl	4	24.100	10.690	2.0961
43	k	2,3,6,4'-tetraCl	4	23.991	10.664	2.1688
44	l	2,4,5,2'-tetraCl	4	24.013	10.762	2.0498
45	m	2,4,5,3'-tetraCl	4	23.832	10.761	2.1202
46	n	2,4,5,4'-tetraCl	4	23.726	10.748	2.1778
47	o	2,4,6,2'-tetraCl	4	24.167	10.675	2.0932
48	p	2,4,6,3'-tetraCl	4	23.984	10.660	2.1805
49	q	2,4,6,4'-tetraCl	4	23.876	10.640	2.2475
50	r	3,4,5,2'-tetraCl	4	23.888	10.974	1.8512
51	s	3,4,5,3'-tetraCl	4	23.711	10.987	1.9047
52	t	3,4,5,4'-tetraCl	4	23.607	10.981	1.9530
53	u	2,3,2',3'-tetraCl	4	24.132	10.802	1.9510
54	v	2,3,2',4'-tetraCl	4	23.996	10.687	2.1409
55	w	2,3,2',5'-tetraCl	4	24.085	10.627	2.1737
56	y	2,3,2',6'-tetraCl	4	24.277	10.676	2.0441
57	z	2,4,2',4'-tetraCl	4	23.861	10.577	2.3251
58	A	2,4,2',5'-tetraCl	4	23.949	10.512	2.3635
59	B	2,4,2',6'-tetraCl	4	24.139	10.553	2.2434
60	C	2,5,2',5'-tetraCl	4	24.038	10.452	2.3963
61	D	2,5,2',6'-tetraCl	4	24.230	10.501	2.2668
62	E	2,6,2',6'-tetraCl	4	24.425	10.563	2.1204
63	F	3,4,3',4'-tetraCl	4	23.579	10.859	2.1032
64	G	3,4,3',5'-tetraCl	4	23.647	10.736	2.2185
65	H	3,5,3',5'-tetraCl	4	23.717	10.618	2.3281

[a] Symbol used in Figs. 6.7–6.10.

the plot, whereas the 3,4,3',4' derivative is at the bottom of a curved area containing all evenly distributed derivatives.

The potential for organizing subsets of molecules from a database is evident from this example. Clearly, more complex structures with greater variation in their structures may be less simply arranged. This illustration reveals the potential for two or three or multidimensional plotting of molecules to reveal similarity or diversity.

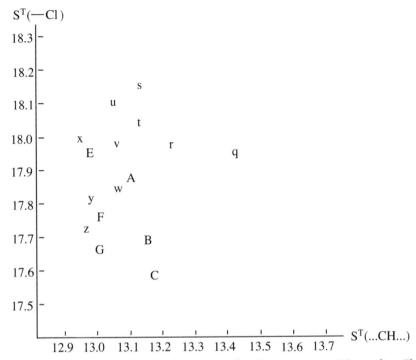

FIGURE 6.10   Plot of trichlorobiphenyls in a space defined by two atom-type E-State indices. The letter codes are defined in the text.

VII. DIVERSITY

In contrast to similarity, the attribute of diversity reflects the differences among a set of structures intended for consideration in an application such as molecule design. This attribute of a set of molecules is of particular value in the design of a library for combinatorial synthesis and automated testing. The intent is to discover lead molecules from a large number of structures. The probability of a discovery of some activity is greatly enhanced if the molecules screened are from a diversity of structures. There may or may not be an active member for a narrow range of structure differences in a set to be tested. The odds are greatly increased if representative structures from a broad range of possibilities are available. This outcome is the benefit of diversity.

A database or a subset of a database may be evaluated for relative diversity using several criteria and techniques. No matter how this is accomplished, the essential common ingredient is the encoding of the molecular structure, especially the encoding of atom and group features within each molecule in a co-

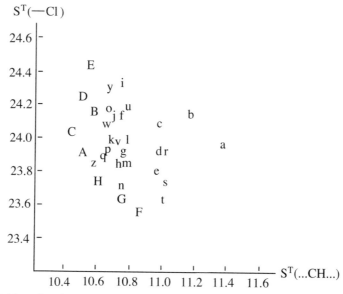

FIGURE 6.11   Plot of tetrachlorobiphenyls in a space defined by two atom-type E-State indices. The letter codes are defined in the text.

herent manner. We have already demonstrated an approach to similarity quantitation using one metric, the Euclidean distance. Using this same metric, it may be possible to encode relative diversity or structure differences among molecules. This is not an easy problem to solve. Current reviews have addressed this problem (Hall *et al.*, 1995; Cummins *et al.*, 1996; Cheng *et al.*, 1996; Lewis *et al.*, 1997; Brown and Martin, 1998; Zheng *et al.*, 1998a,b).

We have found that the E-State paradigm lends itself to database coding, which is useful in both similarity and diversity characterization.

VIII.   VIRTUAL SCREENING

While high-throughput screening (HTS) is making available much larger data sets of synthesized and tested compounds, computational schemes are making it possible to assemble huge computer files of molecular structures. These virtual libraries of structures present significant computational challenges for evaluation to determine those which have significant potential for biological activity or desirable physicochemical properties. Structure representation which is readily computable and easily interpreted is essential to the task. This book demonstrates that these features are two central characteristics of the E-State method.

HTS and combinatorial chemistry (combichem) make available a much greater number of compounds than can be synthesized in practice. Virtual screening is a set of strategies which makes it possible to reduce greatly the actual number needed to be produced. It is necessary for the chemists in combichem projects to find rational methods for deciding which compounds to make and test. As Walters *et al.* (1998) noted, the question is how a chemist should filter the enormous virtual chemistry space. There are computational chemical methods which provide the basis for automatic evaluation of huge chemical libraries. Quantitative structure–activity relationship (QSAR), classification methods, pattern recognition, similarity schemes—based on the E-State—all lend themselves very well to this set of tasks. It should be noted that these QSAR and related methods do not eliminate the need for experience-based chemical intuition; rather, these methods provide aids to intuition which can streamline the decision-making process.

There are at least four stages in the early drug design process in which E-State-based methods may be applied:

1. Database evaluation and construction
2. Similarity analysis
3. QSAR modeling
4. Pharmacophore identification

In the following sections, we present a brief discussion of these issues with respect to the E-State method. This discussion is not exhaustive but is intended to suggest ways to approach these aspects of the overall problem.

A. DATABASE EVALUATION

A critical early step in virtual screening is the characterization of the chemical library. It is important to determine the degree of diversity and/or similarity of the set of structures in the library. This information is the basis for deciding how many compounds need to be considered to create a representative sample for synthesis and testing. Previously, we demonstrated that the E-State indices, including the hydrogen E-State and the atom-type forms, are a sound basis for describing and characterizing chemical structures. Furthermore, methods have been developed to create diverse libraries of specific design. An approach by Zheng *et al.* (1998a,b) makes use of genetic algorithms; this approach uses connectivity chi and kappa indices. Investigations are under way using similar techniques based on the E-State indices.

For characterization of a library, both distance-based and cell-based schemes have been proposed and discussed (Wilson, 1998). It remains to be determined whether one or the other method works best or whether different approaches work best for different problems.

B. SIMILARITY ANALYSIS

Structure representations, such as the E-State indices (Hall *et al.*, 1995) (and molecular connectivity chi and kappa shape indices), have been shown to serve very suitably for similarity analysis (Kier, 1997). The fact that such indices, especially the E-State indices, are computed at very high speed makes them appealing for similarity analysis on very large structure libraries. Various investigators have shown that topological structure representations are often superior to indices based on three-dimensional approaches (Brown and Martin, 1996). The label "2-D," sometimes associated with topological indices, is inaccurate when applied to indices such as the E-State. The E-State is not based on full Cartesian three-dimensional (3-D) geometry and related information; however, this book clearly demonstrates the richness and depth of the structure information which these indices encode. The discussion of 2-D versus 3-D is not particularly useful in this context. The E-State indices (and hydrogen E-State indices) in their atom-type forms encode sufficient information to provide a sound and practical basis for selecting sets of similar compounds in a library (Galvez *et al.*, 1995).

Genetic algorithms (GA) may be used to facilitate compound selection based on topological indices (Brown and Martin, 1998; Zheng *et al.*, 1998a). One problem in the practical use of GA is the development of a sufficiently robust scoring function (Walters *et al.*, 1998). We suggest that similarity (based on atom-type E-State indices) to the appropriate reference or target structure may be a very suitable approach.

Walters *et al.* (1998) point out that the most efficient way to speed up processing of a virtual library is to add more (chemical) intelligence to the process; that is, a structure-based filtering. The atom-type E-State and hydrogen E-State indices provide such a basis. The E-State algorithm is computationally the simplest of the structure-encoding algorithms in use but also a formalism which is rich in structure information. Similarity filtering can be applied to possibilities for synthesis, similarity to known and/or potential drug structures, or similarity to the general structure of a drug class. This added intelligence— structures encoded in the E-State formalism—provides increased speed to the overall process.

C. QSAR ANALYSIS

At some point in the overall modeling process, data sets are produced in a form and size which are appropriate for QSAR analysis. With the typical minimum amount of experimental and structure data needed for modeling, a bootstrap-iterative process can be initiated. The chemist can use the QSAR models to create a list of structures with the greatest predicted potential for good activity;

these are synthesized, tested, and added to the data set. The model can be improved from the updated data set. This iterative process can sharply focus the description of structure features—those which enhance activity and those which are deleterious to activity.

QSAR models based on the E-State (and other topological indices) provide insights into the crucial structure features associated with the activity being modeled. The appearance of atom-level indices in the QSAR equation points to atom-level regions which are significant for activity. The chemist can focus attention on structure modification, associated with the identified atoms, to modify the identified atoms (or nearby regions) to enhance activity. Furthermore, the occurrence of atom-type E-State also suggests the atom types which may be important to the pharmacophore or ancillary anchoring features. The additional possibility of the occurrence of hydrogen E-State indices allows for further refinement of the structure modifications in terms of both polar bonds and lipophilic regions of the structures. Examples of these possibilities are given in Chapters 5 and 7–9.

D. PHARMACOPHORE IDENTIFICATION

Pharmacophore identification remains an important aspect of the overall modeling process. However, it is clear that the topological methods should be used exhaustively to eliminate molecules which cannot match the structure requirements for drug activity. Included in these methods should be topological shape analysis (Galvez et al., 1995). Since the 3-D methods are computationally much more expensive, effective use should first be made of the faster topological methods.

It is clear that topological methods cannot provide a complete specification of a pharmacophore because of the lack of explicit 3-D information. However, topologically based methods do provide some insights into the possible pharmacophore. When atom-level E-State indices occur in a QSAR analysis, the specific atoms identified may suggest possible parts of the pharmacophore or at least indicate small regions of a molecule for consideration. Chapter 7 presents several such studies. We want to make it clear that the specific atoms identified in the topological superposition method are not necessarily actual parts of the putative pharmacophore, but their occurrence in the equation may be highly suggestive. Even the atom-type E-State indices shed some light on possible atom types involved in the pharmacophore.

One of the significant problems in 3-D docking methods is the development of appropriate scoring functions. It was recently suggested that the new E-State field may serve as an efficient and sound basis for a scoring function (G. Maggiora, personal communication).

E. FINAL COMMENTS

Section VIII is intended to suggest new ways for the E-State method (and other topological approaches) to assist in the problems associated with HTS, combi-chem, and virtual libraries. The E-State indices encode much important structure information. As such, these indices provide the basis for a sound chemical structure space. Similarity measures may be applied and the degree of diversity may be assessed by one of several methods. The very same E-State indices and hydrogen E-State indices (and their atom-type forms) may also be used to form QSAR equations for biological activity, toxicity, solubility, or other properties of interest to the overall chemical structure design problem. The combination of very high computability and easy, direct structure interpretation makes the E-State indices very attractive to this array of drug design problems.

IX. EXERCISES

To obtain a sense of the manner in which the E-State values represent molecular structure and facilitate similarity comparison, consider molecules which may be similar in some respects. For each of the molecules, the reader should compute the various E-State indices using the CD-ROM program, E-Calc, including similarity measures such as the Euclidean distance between various molecules based on certain E-State indices—atom level or atom type and E-State and/or hydrogen E-State values. Draw conclusions as to the degree to which the similarity measure indicates the similarity of the molecules.

Exercise 1

Consider *ortho-*, *meta-*, and *para-*chlorophenol. Compute similarity only for the phenolic OH and the chlorine atoms. Extend the similarity measure to other atoms in the molecules. Consider the hydrogen E-State of the phenolic OH group.

Exercise 2

Consider the molecules arising from *ortho* and *meta* substitution on phenol using as substituents the following groups: —CH$_3$, —NH$_2$, —OH, —F, and isopropyl. Consider both the E-State and the hydrogen E-State values for the phenolic OH group as well as for other atoms/groups within the structures.

Exercise 3

Consider molecules of interest to you. Compile lists of E-State and atom-type E-State values for similarity comparisons. Computation of a similarity measure such as the Euclidean distance may not be necessary in order to see the patterns of similarity. Remember that the E-State values are not surrogates for partial charge. They give a measure of electron accessibility for atoms (groups) or electron deficiency for hydrogen E-State values.

X. SUMMARY

The familiar concept of structure similarity is revealed in the set of E-State indices in several ways. Structures arrayed in an E-State space are organized in a structure-coherent fashion. Variations in electron accessibility are revealed as varying positions in the space. Either an average of atom-type E-State index values may be used or the sum of atom-type E-State values for a given molecule may constitute the dimensions of the space. On a coarse scale major groupings of structures are expected as related to counts of atom types. On a fine scale the details of structure variation reveal themselves. The hydrogen E-State indices add information on polar bonds or lipophilic molecular regions. One can take advantage of the overall organization of structures in an E-State space to search for similar compounds or to characterize the diversity of a set of structures. In either case, the computation is very rapid and the results are interpretable in terms of familiar concepts of organic structure. On the basis of the rich information content of the E-State indices and their high-speed computability, the E-State indices offer much to the burgeoning area of combichem and virtual screening.

XI. A LOOK AHEAD

This chapter, which dealt with similarity issues, demonstrates the high level of structure information encoded in the E-State indices. With this insight in mind, we turn our attention to several examples of QSAR analysis. In Chapter 7, the atom-level E-State indices are used to build QSAR models using the topological superposition approach with standard multiple linear regression methods. In subsequent chapters, other indices are joined to the E-State indices, both atom level and atom type, and other statistical methods are employed to develop useful QSAR models of biological and physicochemical properties.

REFERENCES

Brown, R. D., and Martin, Y. C. (1996). Use of structure–activity data to compare structure based clustering methods: Descriptors for use in compound selection. *J. Chem. Inf. Comput. Sci.* **36**, 572–584.

Brown, R. D., and Martin, Y. C. (1998). An evaluation of structural descriptors and clustering methods for use in diversity selection. *SAR QSAR Environ. Res.* **8**, 23–40.

Cheng, C., Maggiora, G., Lajiness, M., and Johnson, M. (1996). Four association constants for relating molecular similarity measures. *J. Chem. Inf. Comput. Sci.* **36**, 909–915.

Cummins, D. J., Andrews, C. W., Bentley, J. A., and Cory, M. (1996). Molecular diversity and chemical databases: Comparison of medicinal chemistry knowledge bases and databases of commercially available compounds. *J. Chem. Inf. Comput. Sci.* **36**, 750–763.

Felder, E., and Poppinger, D. (1997). Combinatorial compound libraries for enhancing drug discovery approaches. *Adv. Drug Res.* **30**, 111–199.

Galvez, J., Garcia-Domenech, R., de Julian-Ortiz, J. V., and Soler, R. (1995). Topological approach to drug design. *J. Chem. Inf. Comput. Sci.* **35**, 272–284.

Hall, L. H., and Kier, L. B. (1995). Electrotopological state indices for atom types. *J. Chem. Inf. Comput. Sci.* **35**, 1039–1045.

Hall, L. H., Kier, L. B., and Brown, B. B. (1995). Molecular similarity based on novel atom-type E-State indices. *J. Chem. Inf. Comput. Sci.* **35**, 1074–1080.

Johnson, M., and Maggiora, G. (Eds.) (1990). *Concepts and Applications of Molecular Similarity.* Wiley-Interscience, New York.

Kier, L. B. (1997). Kappa shape indices for similarity analysis. *Med. Chem. Res.* **7**, 394–406.

Lewis, R. A., Mason, J. S., and McLay, I. M. (1997). Similarity measures for rational set selection and analysis of combinatorial libraries: The diverse property-derived (DPD) approach. *J. Chem. Inf. Comput. Sci.* **37**, 599–614.

Testa, B., and Kier, L. B. (1991). The concept of structure in structure–activity relations by studies and drug design. *Med. Res. Rev.* **11**, 35–48.

van Drie, J., and Lajiness, M. (1998). Approaches to virtual library design. *Drug Discovery Topics* **3**, 274–283.

Walters, W. P., Stahl, M. T., and Murcko, M. (1998). Virtual screening—An overview. *Drug Discovery Topics* **3**, 260–278.

Wilson, E. K. (1998). Computers customize combinatorial libraries. *C & E News* **76**, 31–37.

Zheng, W., Cho, S. J., and Tropsha, A. (1998a). Rational combinatorial design. 1. Focus 2-D: A new approach to the design of targeted combinatorial chemical libraries. *J. Chem. Inf. Comput. Sci.* **38**, 251–258.

Zheng, W., Cho, S. J., and Tropsha, A. (1998b). Rational combinatorial design. 2. Rational design of targeted combinatorial peptide libraries using chemical similarity probe and inverse QSAR approaches. *J. Chem. Inf. Comput. Sci.* **38**, 259–268.

CHAPTER 7

Structure–Activity Studies: Atom-Level Indices

I. THE STRUCTURE–ACTIVITY PARADIGM

The E-State indices form a pattern of structure descriptors encoding the electron richness and topological availability of atoms within a molecule. These attributes are related to molecular properties, arising from interactions among a set of identical molecules and interactions among different molecules. The measurement of properties provides numerical values that reflect the molecular structure. Changes in structure within a series of molecules produce changes in measured property values, with the relationship constituting an equation if a structure can be quantified. This quantification as molecular structure representation has been the subject of previous chapters in which the E-State index was derived and explained. In this chapter, several examples of structure–activity analyses using the E-State indices are presented in which only the atom-level E-State indices are used. In these analyses, we also adopt a topological superposition method.

II. INHIBITION OF MONOAMINE OXIDASE BY HYDRAZIDES

Monoamine oxidase (MAO) is an important enzyme in the metabolism of several neurotransmitters, including dopamine, serotonin, and norepinephrine. In cases of deficiencies of any of these, the inhibition of the metabolizing enzyme leads to the preservation of the transmitter, producing a clinical effect equivalent to that of their direct use. The search for effective inhibitors is an objective of drug design. Data from Fulcrand *et al.* (1978) is available for a set of 24 aryloxyacetohydrazide derivatives tested for their inhibition of MAO. The potency is expressed as the concentration needed to cause 50% inhibition, converted to the negative logarithm $-\log(IC_{50}) = pIC_{50}$ for analysis. In this data set each molecule possesses atoms in the same 14 molecular skeletal positions, as shown in Scheme I.

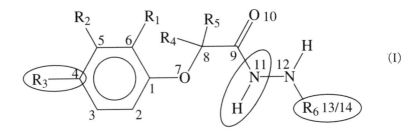

(I)

The E-State index was computed for each atom in each molecule and those for the 14 common skeletal atoms were entered into regression analysis with the potency. Equations were obtained for all possible one-, two-, and three-variable models, and the best model in each case was recorded. The best one-variable model yielded the following correlation coefficient and standard deviation: $r = 0.66$, $s = 0.42$. This model was based on the E-State value for the atom No. 1 in Scheme I, a carbon atom in substituent R_6 (Table 7.1). The best two-variable equation is a statistically significant model:

$$pIC_{50} = -1.77 \ (\pm 0.15) \ S(C_1) - 9.42 \ (\pm 1.08) \ S(NH_5) \qquad (7.1)$$
$$+ \ 32.76 \ (\pm 2.92)$$

$$r^2 = 0.88, \ s = 0.20, \ F = 76, \ n = 24$$

$$r^2_{press} = 0.83, \ s_{press} = 0.24$$

The press statistics are based on the leave-one-out cross-validation method. The following is the best three-variable model:

$$pIC_{50} = -1.69 \ (\pm 0.14) \ S(C_1) - 9.14 \ (\pm 1.01) \ S(NH_5) \qquad (7.2)$$
$$+ \ 0.15 \ (\pm 0.07) \ S(C_{13}) + 32.15 \ (\pm 2.72)$$

$$r^2 = 0.90, \ s = 0.19, \ F = 61, \ n = 24$$

$$r^2_{press} = 0.85, \ s_{press} = 0.23$$

The data set is shown in Table 7.1 along with the predicted values. The symbol $S(NH\text{-}11)$ refers to the E-State index value for the No. 5 group, NH, adjacent to the carbonyl group in Scheme I. $S(C\text{-}4)$ refers to the E-State index value for the phenyl ring carbon atom (No. 4 group in Scheme I) para to the ether link. In the three-variable equation the $S(NH\text{-}11)$ and $S(C\text{-}14)$ variables combine to account for 90% of the calculated activity, suggesting that primary efforts in drug design might be focused in this area.

The press statistics indicate that these models may be used for predictive purposes since the press standard deviation diminishes by such a small amount

TABLE 7.1 Hydrazide Monoamine Oxidase Inhibitors

		Substituents in Scheme I				$S(C\text{-}14)$	$S(NH\text{-}11)$	pIC_{50}	Calcu-lated	Re-sidual
R_1	R_2	R_3	R_4	R_5	R_6					
H	H	H	H	H	$CH(CH_3)_2$	1.943	2.538	5.42	5.42	0.00
Cl	H	H	H	H	$CH(CH_3)_2$	1.926	2.515	5.60	5.66	−0.06
H	Cl	H	H	H	$CH(CH_3)_2$	1.929	2.520	5.40	5.61	−0.21
H	H	Cl	H	H	$CH(CH_3)_2$	1.931	2.524	5.96	5.57	0.39
CH_3	H	H	H	H	$CH(CH_3)_2$	1.947	2.558	5.54	5.22	0.32
H	CH_3	H	H	H	$CH(CH_3)_2$	1.946	2.553	5.05	5.27	−0.22
H	H	CH_3	H	H	$CH(CH_3)_2$	1.946	2.550	5.40	5.30	0.10
OCH_3	H	H	H	H	$CH(CH_3)_2$	1.932	2.535	5.62	5.47	0.15
H	OCH_3	H	H	H	$CH(CH_3)_2$	1.934	2.536	5.42	5.45	−0.03
H	H	OCH_3	H	H	$CH(CH_3)_2$	1.935	2.536	5.52	5.45	0.07
H	H	H	H	CH_3	$CH(CH_3)_2$	1.947	2.588	5.00	4.94	0.06
Cl	H	H	H	CH_3	$CH(CH_3)_2$	1.930	2.565	5.16	5.19	−0.03
H	Cl	H	H	CH_3	$CH(CH_3)_2$	1.933	2.570	4.96	5.13	−0.17
H	H	Cl	H	CH_3	$CH(CH_3)_2$	1.936	2.573	5.00	5.10	−0.10
H	H	H	CH_3	CH_3	$CH(CH_3)_2$	1.950	2.629	4.34	4.55	−0.21
H	H	Cl	CH_3	CH_3	$CH(CH_3)_2$	1.938	2.614	4.80	4.71	0.09
H	CH_3	H	H	CH_3	$CH(CH_3)_2$	1.951	2.603	4.90	4.79	0.11
H	H	H	H	H	C_2H_5	1.901	2.489	5.82	5.95	−0.07
H	H	Cl	H	H	C_2H_5	1.889	2.474	6.00	6.12	−0.12
H	H	H	H	H	$CH_2C_6H_5$	1.103	2.587	6.14	6.44	−0.30
H	H	H	H	H	$CH(CH_3)C_6H_5$	1.108	2.637	5.70	5.96	−0.26
H	H	CH_3	H	H	$CH(CH_3)C_6H_5$	1.109	2.649	6.05	5.85	0.15
H	H	OCH_3	H	H	$CH(CH_3)C_6H_5$	1.096	2.635	6.00	6.00	0.00
H	H	Cl	H	H	$CH_2C_6H_5$	1.089	2.573	6.96	6.60	0.36

from the regression standard deviation. The previous equations indicate that the potency of the hydrazide derivative can be successfully modeled with the two or three atom E-State indices. It should be noted that the C-14 atom stands for four kinds of carbon atoms: methyl groups in either an ethyl (compounds 18 and 19) or isopropyl group (compounds 1–17), the aromatic carbon which is directly attached to the methylene of a benzyl group (compounds 20 and 24), or the aromatic carbon in the α-methyl benzyl group (compounds 21–23). These models indicate that there are three parts of these molecules which are important for inhibition since the E-State values at these three points parallel the variation in the inhibitory power.

Investigation of these data was extended to include analyses using molecular orbital theory. One motive in this study was interest in the possible correlation between E-State values and computed partial charges. Although the E-State formalism does not suggest a significant relationship as described in an earlier chapter, some researchers were concerned with this. Second, there was a desire to study a possible correlation with values of the highest occupied molecular orbital (HOMO) and the lowest unoccupied molecular orbital (LUMO) energies. Conformational analysis was carried out using the AM1 Hamiltonian. For structures obtained in the minimum-energy conformation, partial charges were calculated along with HOMO and LUMO energies. For atoms with attached hydrogen atoms, the partial charge for the hydrogen(s) was added to that of the nonhydrogen atom to obtain a charge, Tq_i. Correlations between the E-State index and the corresponding Tq_i indicated that only at position 1 (C1) was there significant correlation. In addition, one other correlation coefficient was 0.80 and all others were less than 0.45. For the entire set of molecules, there appears to be no general correlation between E-State values and computed partial charges.

The Tq_i values were evaluated by the same statistical analysis used for the 14 E-State indices. The best two- and three-variable equations are as follows:

$$pIC_{50} = -11.1 \ (\pm 1.63) \ Tq_{14} - 62.1 \ (\pm 15.4) \ Tq_{11} + 3.64 \ (\pm 0.48) \quad (7.3)$$

$$r^2 = 0.70, \ s = 0.32, \ F = 25, \ n = 24$$

$$r^2_{press} = 0.58, \ s_{press} = 0.38$$

$$pIC_{50} = -11.3 \ (\pm 1.37) \ Tq_{14} - 83.9 \ (\pm 14.7) \ Tq_{11} \quad (7.4)$$
$$+ \ 29.2 \ (\pm 9.34) \ Tq_9 + 4.59 \ (\pm 2.66)$$

$$r^2 = 0.80, \ s = 0.27, \ F = 26, \ n = 24$$

$$r^2_{press} = 0.70, \ s_{press} = 0.33$$

Two general observations can be made from these results. First, the quality of these charge-based models is significantly less than that obtained with the E-State indices. Both the direct regression statistics and the press statistics are significantly inferior and so the models lack predictive value. In particular, the press standard errors are 50% greater than those obtained with the E-State indices. Second, the two most important MO variables refer to positions 14 and 11, the same as those in the E-State models; also, the MO-based models suggest the same areas for attention in drug design. However, the time and cost of creating the E-State model is a tiny fraction of that required for the MO model and the resulting E-State model is significantly better. The high quality of the quantitative structure–activity relationship (QSAR) modeling for this data set supports the usefulness of this topological superposition strategy. Furthermore, it suggests that the electron accessibility information encoded in the E-State indices is a useful representation of molecular structure. In the following sections, several other studies are discussed that illustrate the application of atom-level E-State indices.

III. ODOR THRESHOLD OF ALKYLPYRAZINES

A hypothesis put forward by Amoore (1967) states that the odor characteristics of molecules are functions of their size and shape. A number of examples over the years have lent support to this idea. An extension of this hypothesis might address the attributes of a molecule from the perspective of the molecular topology. In addition, it is hard to exclude consideration of the electronic structure from a list of molecular attributes influencing olfaction. With these possibilities in mind, a study by Tsantili-Kakoulidou and Kier (1992) examined structure indices encoding atom connectivity, size, shape, and electron richness of a series of odorous alkylpyrazines (Scheme II). Among these indices were the E-State values of atoms in the molecules which proved to be useful in creating a structure–activity model of the odor property.

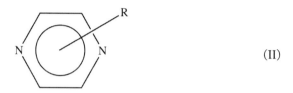

(II)

The odor thresholds were defined as the vapor concentration [parts per billion (ppb)] necessary to evoke a recognition of odor for each molecule in the set (Teranishi *et al.*, 1974). In this study, the threshold concentrations, expressed as log ppb, for 13 un-, mono-, di-, and trisubstituted alkylpyrazines are available (Scheme II). The E-State indices were calculated for each atom of the ring, whereas the E-State values for the nitrogen atoms were calculated and averaged, using this as a single index $S(N)$. All the indices were regressed against the log ppb values. The best single variable equation was found to be

$$\log ppb = -10.49 \ S(N) + 45.09 \tag{7.5}$$

$$r^2 = 0.931, \ s = 0.434, \ n = 13$$

This relationship was found to be superior to all other topological parameters tested in the study (Tsantili-Kakoulidou and Kier, 1992). A plot of the variables in Eq. (7.5) revealed a curvature, suggesting that a quadratic expression in $S(N)$ might improve the model. This was found to be true as seen in Eq. (7.6):

$$\log ppb = 94.87 \ S(N) - 13.17 \ [S(N)]^2 - 165.38 \tag{7.6}$$

$$r^2 = 0.982, \ s = 0.252, \ n = 13$$

148

TABLE 7.2   Odor Threshold of Alkylpyrazines

No.	Substituent	$S(N)$	Observed log ppb	Calculated log ppb	Residual
1	2,5-Dimethyl, 4-ethyl	4.258	−0.40	−0.15	−0.25
2	2,6-Diethyl	4.206	0.78	0.72	0.06
3	2,5-Diethyl	4.206	1.30	0.72	0.58
4	2-Methyl, 3-isobutyl	4.126	1.54	1.90	−0.36
5	2-Methyl, 5-ethyl	4.111	2.00	2.11	−0.11
6	2-Methyl, 3-ethyl	4.111	2.11	2.11	0.00
7	2-Isobutyl	4.066	2.60	2.69	−0.09
8	2,6-Dimethyl	4.016	3.18	3.26	−0.08
9	2,5-Dimethyl	4.016	3.26	3.26	0.00
10	2,3-Dimethyl	4.016	3.40	3.26	0.14
11	2-Ethyl	3.964	3.78	3.79	−0.01
12	2-Methyl	3.870	4.78	4.57	0.21
13	H	3.722	5.24	5.32	−0.08

The experimental and predicted values of log ppb are provided in Table 7.2. Equation (7.6) reveals a prominent role for the nitrogen atoms of pyrazine, influencing the odor threshold. It also reveals that the electronic structures of these atoms are important since their local topology is constant in this series. Larger values for the nitrogen E-State indicate greater electron richness at that atom. High values of the E-State for the nitrogens correspond to low odor threshold. High values for $S(N)$ correlate with low values for log ppb (low potency); that is, high odor threshold. This may arise from an enhanced binding at a receptor through the lone-pair electrons on the nitrogens, which produce electrostatic binding or hydrogen bonding with a receptor feature. The quadratic model indicates that there is an optimum value of about 3.60 for the E-State index on the nitrogen atoms. The E-State indices produce a very good correlation in this study and make it possible to draw inferences about structure influencing the properties in this series.

IV. XANTHINE INHIBITORS OF ADENOSINE A_1

Considerable interest has been shown in purine analogs as potential agents effective in the nervous and cardiovascular systems (Stone, 1989). One focus of this research has been the interactions of several classes of purines with different receptors. Various methods have been used to study the possible binding modes leading to several hypotheses based on molecular electrostatic potential maps (Van Galen *et al.*, 1990; Peet *et al.*, 1990). With the use of the E-State method, it may be possible to explore the salient electronic features influencing the critical binding modes of these molecules. Such a study was reported by Joshi and Kier (1992), who considered a series of 1,3 dialkyl xanthines (Scheme III), reported to be selective antagonists at A_1-adenosine receptors (Jacobsen *et al.*, 1989). One subset of compounds in the inhibition analysis contained large polar extensions on the 8-phenyl ring, a modification which might confound an attempt to analyze the important atoms in the xanthine ring system. These compounds were excluded in the structure–activity study of Joshi and Kier (1992).

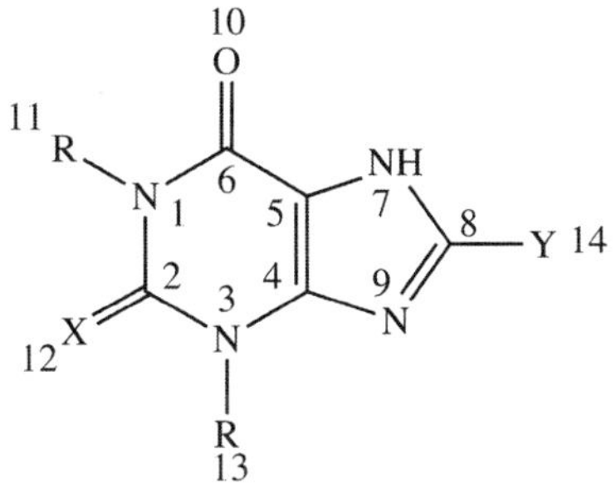

(III)

The 28 compounds used in the study (Table 7.3) were analyzed for their E-State values. Two tautomers can exist for the xanthine molecule involving the N(7) and N(9) atoms shown in Scheme III. For most substituted xanthines the N(7)–H tautomer predominates (Elguero *et al.*, 1976), and this was adopted by Joshi and Kier (1992). The first atom of the substituent on position 8 was given the identity number 14. The activity of the xanthines was expressed as the affinity, log K_i, for the A_1-adenosine receptor. The best two-variable equation was found to include the E-State values for the N(7) and the keto oxygen atom O(10) with an r^2 of 0.84. Two three-variable equations were found to be of equal quality:

$$\log K_i = -1.17 \, S(\text{NH-7}) - 0.97 \, S(\text{O-10})$$

$$- 0.22 \, S(\text{X-12}) + 1.71 \quad (7.7)$$

$$r^2 = 0.88, \, s = 0.33, \, F = 61, \, n = 28$$

$$\log K_i = -1.07 \, S(\text{NH-7}) - 0.91 \, S(\text{O-10})$$

$$- 0.19 \, S(\text{Y-14}) + 1.71 \quad (7.8)$$

$$r^2 = 0.88, \, s = 0.34, \, F = 59, \, n = 28$$

A four-variable equation combines these parameters into an equation explaining 93% of the variation in the data:

$$\log K_i = -1.21\ S(\text{NH-7}) - 1.02\ S(\text{O-10}) - 0.23\ S(\text{X-12}) \tag{7.9}$$
$$+\ 0.20\ S(\text{Y-14}) + 1.71$$

$$r^2 = 0.93,\ s = 0.26,\ F = 77,\ n = 28$$

The calculated values using Eq. (7.9) are found in Table 7.3. The equation models the fact that activity is greater when oxygen rather than sulfur is in position 12. Structure changes producing greater electron richness in either oxygen atom enhance activity. The nitrogen atoms in positions 1 and 3 do not contribute significantly, whereas the nitrogen at position 7 is important. Whether this tautomer or the alternative one is present is not clear, but

TABLE 7.3   Adenosine Inhibition by Xanthine Derivatives

R	X	Y	$\log K_i$	Calculated $\log K_i$	Residual
Me	O	H	3.93	3.58	0.35
Pr	O	H	2.65	2.97	−0.32
Me	O	Cyclopentyl	1.04	0.96	0.08
Pr	O	Cyclopentyl	−0.05	−0.06	0.01
Me	S	Cyclopentyl	1.01	0.83	0.18
Pr	S	Cyclopentyl	−0.18	−0.16	−0.02
Me	S	Cyclopentyl	1.61	1.93	−0.32
Pr	S	Cyclopentyl	0.69	1.07	−0.38
Me	O	Cyclopentyl	2.31	2.06	0.25
Pr	O	Cyclopentyl	1.19	1.17	0.02
Me	O	2-Furyl	2.54	2.58	−0.04
Pr	O	2-Furyl	1.57	1.58	−0.01
Me	S	2-Furyl	2.26	2.45	−0.19
Pr	S	2-Furyl	1.51	1.49	0.02
Me	O	3-Furyl	1.86	2.52	−0.66
Me	O	2-Thienyl	2.37	2.22	0.15
Me	O	3-Thienyl	2.18	2.22	−0.04
Pr	O	2-Thienyl	1.21	1.22	−0.01
Pr	O	3-Thienyl	1.00	1.23	−0.23
Me	S	2-Thienyl	2.34	2.09	0.25
Pr	S	2-Thienyl	1.55	1.12	0.43
Me	O	Phenyl	1.93	1.87	0.06
Et	O	Phenyl	1.65	1.54	0.11
Pr	O	Phenyl	1.01	1.12	−0.11
Me	S	Phenyl	1.58	1.75	−0.17
Pr	S	Phenyl	1.21	1.03	0.18
Me	O	Phenyl	3.14	2.99	0.15
Et	O	Phenyl	3.00	2.73	0.27

the study does reveal a close correlation between $S(NH\text{-}7)$ and $S(N\text{-}9)$. Based on this model the five-membered ring, including the 14 substituent, is predicted to be a strong contributor to the binding event.

V. BENZIMIDAZOLES AS INFLUENZA VIRUS INHIBITORS

Benzimidazoles (Scheme IV) have been reported for their effect in inhibiting influenza viruses (Tamm *et al.*, 1953). Hall *et al.* (1991a) studied these compounds, calculating the E-State indices and comparing these with the inhibition constants, pK_i. The tautomerism encountered between the two nitrogen atoms in Scheme IV was handled by using an average intrinsic state value, I, for each nitrogen atom derived from the values $I(\text{—NH—}) = 2.50$ and $I(\text{=N—}) = 3.00$; hence, $I(N_{1,3}) = 2.75$. A multiple linear regression analysis produced the following equation:

$$pK_i = 4.27\ (\pm0.44)\ S(N\text{-}1,3) + 0.79\ (\pm0.19)\ S(C\text{-}2) \tag{7.10}$$
$$- 16.00\ (\pm2.04)$$

$$r^2 = 0.91,\ s = 0.16,\ F = 63,\ n = 15$$

(IV)

The experimental and calculated values are shown in Table 7.4. These results implicate the five-membered ring as playing a significant role at the receptor, in agreement with results from an earlier structure–activity study (Hall and Kier, 1978). The robustness of the equation was demonstrated by using a leave-one-out method to test the prediction statistics of the remaining compounds (Table 7.5). These statistical results indicate the stability of the model when each compound is left out and the analysis is redone. The coefficients of the two E-State variables do not differ significantly from the standard regression equation. Likewise, the average residual (mean absolute error) is 0.12 for the regression analysis and only slightly higher (0.15) for the leave-one-out analysis.

Inspection of the data shows that for the relation between activity and the E-State index for nitrogens, $S(N\text{-}1,3)$, there are two groups of compounds:

TABLE 7.4   Benzimidazoles Inhibiting Influenza Viruses

Substituent	$S(N-1,3)$	$S(C-2)$	pK_i	Calculated	Residual
Benzimidazole	4.008	1.574	2.14	2.35	−0.21
2-Methyl	4.202	0.848	2.51	2.60	−0.09
5-Methyl	4.059	1.587	2.72	2.57	0.15
5,6-Dimethyl	4.110	1.601	2.72	2.80	−0.08
4,6-Dimethyl	4.139	1.608	2.82	2.93	−0.11
2,5-Dimethyl	4.253	0.852	2.89	2.82	0.07
4,5-Dimethyl	4.139	1.608	2.96	2.93	0.03
2,5,6-Trimethyl	4.304	0.856	3.05	3.04	0.01
2,4,5-Trimethyl	4.333	0.860	3.20	3.17	0.03
5,6-Diethyl	4.192	1.629	3.39	3.17	0.22
2-Propyl, 5-methyl	4.442	0.975	3.60	3.73	−0.13
2,4,5,6,7-Pentamethyl	4.464	0.871	3.66	3.74	−0.08
2-Ethyl-5-methyl	4.371	0.940	3.74	3.40	0.34
2-Butyl-5-methyl	4.490	0.993	3.77	3.95	−0.18
2-Isopropyl-5-methyl	4.452	0.945	3.77	3.75	0.02

those which are substituted at the 2 position and those that are not. Correlation with the $S(N-1,3)$ variable yields $r = 0.794$ and indicates the two distinct groups of compounds. When the two-variable equation is examined, it is clear that the combination of the two variables, $S(N-1,3)$ and $S(C-2)$, accounts for the two groups of compounds. The effect of the substituents on the six-membered ring is indirect but well encoded in the two variables.

VI. BINDING OF BARBITURATES TO CYCLODEXTRIN

A series of barbiturates (Scheme V) were evaluated for their ability to bind to α- and β-cyclodextrin molecules (Scheme VI) (Uekama *et al.*, 1978).

TABLE 7.5   Summary of Leave-One-Out Method of Benzimidazole Data

Quantity	From regression	From leave-one-out method
$S(N-1,3)$	4.28 (0.44)	4.29 (0.15)
$S(C-2)$	0.786 (0.19)	0.790 (0.05)
Intercept	−16.00 (2.04)	−16.08 (0.68)
Average residual	0.116	0.145

Note. Numbers in parentheses are standard deviations.

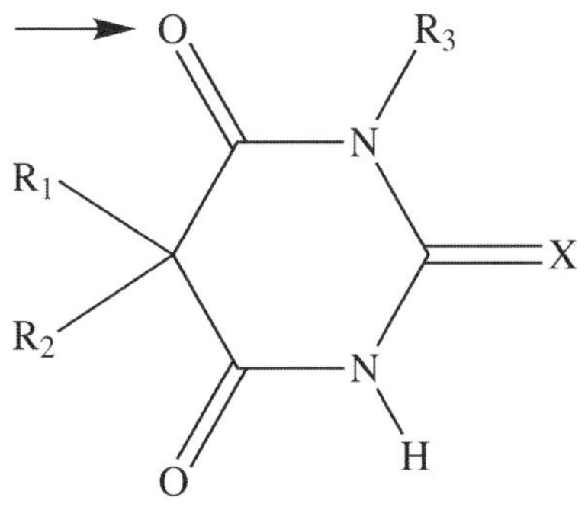

(V)

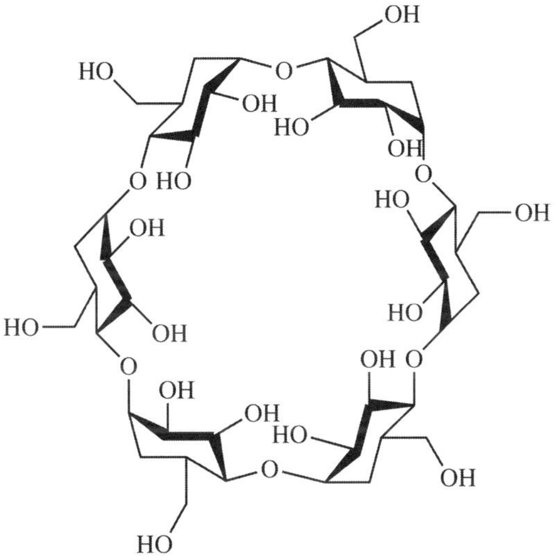

(VI)

Some structure–activity analyses have used indicator variables and molar re-
fraction (Lopata *et al.*, 1985) and kappa shape indices (Kier, 1986). Both
of these studies considered the structure of the barbiturates through whole-
molecule descriptors. As a result, the information about structure influencing
the binding to cyclodextrin was relevant only to the side chain since it is the
major position of structure variation in the series. These data have been analyzed by Kier and Hall (1990) using the E-State indices for both the ring and
the attached atoms. This investigation provided an opportunity to determine
the possible role of the barbiturate ring system in the binding encounter. In this
study, both the α- and β-cyclodextrin molecules were analyzed for binding to
the set of barbiturates. Binding was quantified as the logarithm of the stability
constant, log K_i. No significant correlation was found between a barbiturate
ring atom and binding to an α-cyclodextrin molecule. In contrast, a very good

relationship was found for binding to β-cyclodextrin, expressed as $\log K_a$. The E-State index found to be important corresponds to the oxygen of the carbonyl group, $>C{=}O$. The following is the equation for the QSAR model:

$$\log K_i = 101.8\, S({=}O) - 4.21\, S({=}O)^2 - 612.2 \tag{7.11}$$

$$r^2 = 0.81, \; s = 0.20, \; F = 28, \; n = 16$$

The predicted values are shown in Table 7.6. The structure–activity analysis with the E-State values implicates the carbonyl oxygen atom shown by an arrow in Scheme V. The quadratic equation indicates that there is an optimum E-State value of this atom corresponding to a maximum binding stability for this set of compounds with β-cyclodextrin. This optimum corresponds to an E-State value of $S({=}O) = 12.10$. The predicted optimum value for the binding is $\log K_i = 3.72$, which is very close to that for compound 14 in the data set, the compound with the highest experimental binding. These results closely correspond to those reported earlier for the β-cyclodextrin molecule (Lopata *et al.*, 1985) in which a significant part of the binding strength was attributed to a part of the ring. We identify this ring feature as the carbonyl oxygen atom. The results indicate that the E-State index of this position, which reveals strong influence of the X and R_3 substituents, does indeed parallel the binding. It could be reasoned that this oxygen atom is involved in a hydrogen bond with one or more of the hydroxyl groups around the rim of the β-cyclodextrin molecule.

TABLE 7.6   Binding of Barbiturates to β-Cyclodextrin

Structure variations in Scheme IV					$\log K_i$	
R_1	R_2	R_3	X	$S({=}O)$	Observed	Calculated
$-CH_2CH_2CH_3$	$-CH_2CH_3$	H	O	11.57	2.11	2.28
$-CH_2CH_2CH_2CH_3$	$-CH_2CH_3$	H	O	11.68	2.68	2.74
$-CH_2CH_2CH_2CH_2CH_3$	$-CH_2CH_3$	H	O	11.77	3.11	3.01
$-CH_2CH_2CH_2CH_2CH_2CH_3$	$-CH_2CH_3$	H	O	11.84	3.46	3.18
$-CH(CH_3){-}CH_2CH_2CH_3$	$-CH_2CH_3$	H	O	11.92	3.20	3.31
$-CH_2CH_2CH(CH_3)_2$	$-CH_2CH_3$	H	O	11.79	3.24	3.06
$-$Phenyl	$-CH_2CH_3$	H	O	11.98	3.27	3.39
$-$Phenyl	$-CH_2CH_3$	CH_3	O	12.34	3.22	3.17
$-$Cyclohex-3-enyl	$-CH_3$	CH_3	O	12.19	3.19	3.40
$-CH_2CH_3$	$-CH_2CH_3$	H	S	11.57	2.48	2.29
$-CH_2CH_2CH_3$	$-CH_2CH_3$	H	S	11.73	2.73	2.88
$-CH_2CH_2CH_2CH_3$	$-CH_2CH_3$	H	S	11.84	2.84	3.18
$-CH_2CH_2CH_2CH_2CH_3$	$-CH_2CH_3$	H	S	11.93	3.32	3.33
$-CH_2CH_2CH_2CH_2CH_2CH_3$	$-CH_2CH_3$	H	S	12.00	3.68	3.40
$-CH(CH_3){-}CH_2CH_2CH_3$	$-CH_2CH_3$	H	S	12.05	3.38	3.43
$-$Phenyl	$-CH_2CH_3$	H	S	12.14	3.55	3.43

It is clear that the E-State analysis of this system has identified one structure feature that may be manipulated to modify the binding with cyclodextrin. The structure–activity analysis proceeded in a straightforward manner to identify the important region of the molecules.

VII. IMIDAZOLE INHIBITORS OF THROMBOXANE SYNTHASE

Thromboxane synthase is an enzyme in the arachidonic acid pathway responsible for the synthesis of thromboxane A_2. This molecule is a potent inducer of blood platelet aggregation and vasoconstriction. Its effect is balanced by the opposite effects of prostacyclin, leading to a normal hemodynamic state in the body. An imbalance leading to the dominance of thromboxane produces several pathophysiological conditions, including thrombosis, ischemia, mycardial infarction, and renal failure. It is desirable to inhibit production of thromboxane when out of balance, which may be accomplished by inhibiting its synthesis, most directly by inhibiting thromboxane synthase. Several imidazole derivatives (Scheme VII) have been synthesized and tested as inhibitors (Iizuka *et al.*, 1981).

(VII)

Burnett (1998) analyzed compounds which contain carboxyl groups using E-State indices. The following is the best two-variable equation relating the inhibitory potency, $-\log IC_{50}$:

$$-\log IC_{50} = 1.51\ S(C_{carboxyl}) - 1.41\ S(N\text{-}1) + 11.45 \qquad (7.12)$$

$$r^2 = 0.93,\ s = 0.33,\ F = 86,\ n = 15$$

The predicted values are shown in Table 7.7. The equation model reveals an important role for the carboxyl group and a ring nitrogen atom. A higher E-State value on the carboxyl carbon atom is associated with greater potency in the inhibition of the synthase enzyme. The nitrogen atom, designated No. 1 in Scheme VII, contributes to the potency when it is substituted and when this substituent has electron-withdrawing features. These conclusions, derived from a model using E-State indices, are useful in guiding further modifications of this series of compounds.

TABLE 7.7   Imidazole Inhibitors of Thromboxane Synthase

Position and group in Scheme VII	$S(N)$	$S(C)$	Experimental log IC_{50}	Calculated log IC_{50}
$2\text{-}CH_2(CH_2)_5COOH$	4.113	−0.698	4.69	4.59
$5\text{-}CH_2(CH_2)_5COOH$	3.927	−0.696	4.69	4.89
$4\text{-}CH_2(CH_2)_5COOH$	4.113	−0.697	4.69	4.59
$1\text{-}CH_2(CH_2)_5COOH$	2.039	−0.695	7.41	7.51
$1\text{-}CH_2(CH_2)_5CH(CH_2OH)COOH$	2.035	−0.897	6.60	7.20
$1\text{-}CH_2(CH_2)_6COOH$	2.067	−0.687	7.49	7.49
$1\text{-}CH_2(CH_2)_5CH(COOH)COOH$	1.999	−1.532	6.26	6.36
$1\text{-}CH_2C\equiv C(CH_2)_4COOH$	1.897	−0.735	7.52	7.62
$1\text{-}CH_2C\equiv C(CH_2)_3C(CH_3)_2COOH$	1.897	−0.746	8.05	7.65
$1\text{-}CH_2(CH_2)_5CH(CH_3)COOH$	2.069	−0.682	7.52	7.52
$1\text{-}CH_2(CH_2)_5C(CH_3)_2COOH$	2.044	−0.706	7.52	7.52
$1\text{-}CH_2(CH_2)_5C(=CH_2)COOH$	2.052	−0.877	7.80	7.20
$1\text{-}C(CH_3)_2CH_2(CH_2)_5COOH$	2.129	−0.691	7.04	7.44
$1\text{-}CH_2(CH_2)_5CH(Cl)COOH$	2.043	−0.917	7.60	7.20
$1\text{-}CH_2(CH_2)_5CH(OH)COOH$	2.020	−1.120	6.85	6.95

VIII. BINDING OF SALICYLAMIDES

Several 2-pyrrolidinylmethyl salicylamides (Scheme VIII) have been reported
as possible antipsychotic agents (Ogren *et al.*, 1986; Hogberg *et al.*, 1987).

(VIII)

These compounds exhibit strong binding affinity to the dopamine D_2 receptor.
Structure–activity studies based on physical properties have provided limited
information on the importance of certain structure features. The use of the
E-State indices is one approach to the identification of the parts of these molecules that influence the binding strength. Such a study was reported in which
the binding of several salicylamides, expressed as the inhibition of the binding

of the substrate [^{3}H]spiperone, was analyzed with E-State values of atoms in these molecules (Hall and Kier, 1992). The potency, expressed as $pIC_{50} = -\log[IC_{50}]$, was determined to correlate well with two E-State indices:

$$pIC_{50} = 4.38\ (\pm 0.34)\ S(7) - 2.55\ (\pm 0.50)\ S(8) - 29.26\ (\pm 2.83)(7.13)$$

$$r^2 = 0.88,\ s = 0.24,\ F = 88,\ n = 25$$

$$r^2_{press} = 0.87,\ r^2 = 0.27$$

The predicted potencies are shown in Table 7.8. The intercorrelation of the two variables is very low at $r^2 = 0.37$. The fact that the press statistics show very little decrease in statistical significance suggests that this relation can be used for prediction with confidence. The E-State indices in this equation model are those of the hydroxyl group and the methoxy oxygen atom on the benzene

TABLE 7.8   Salicylamide Inhibition of [^{3}H]Spiperone Binding

Groups in Scheme VIII							Predicted
R_3	R_5	$S(7)$	$S(8)$	pIC_{50}	Calculated	Residual	residual
Cl	Cl	10.05	5.13	1.59	1.68	−0.09	−0.10
Cl	Et	10.21	5.35	1.92	1.79	0.13	0.15
Br	F	10.03	4.92	2.18	2.10	0.08	0.11
Br	Et	10.29	5.39	1.77	2.06	−0.29	−0.33
Et	Cl	10.38	5.23	3.05	2.83	0.22	0.25
Et	Br	10.41	5.31	2.74	2.78	−0.04	−0.05
Et	Et	10.53	5.45	2.92	2.94	−0.02	−0.02
Me	Me	10.24	5.32	2.11	2.03	0.08	0.09
Me	Cl	10.18	5.18	2.59	2.11	0.48	0.51
Br	H	10.07	5.16	1.92	1.66	0.26	0.27
Cl	H	9.99	5.13	1.41	1.39	0.02	0.02
H	H	9.83	5.13	0.50	0.70	−0.20	−0.24
Me	H	10.12	5.19	1.72	1.82	−0.10	−0.10
Et	H	10.31	5.23	2.33	2.54	−0.21	−0.23
H	Br	9.94	5.21	1.25	0.94	0.31	0.35
H	Cl	9.90	5.13	1.19	0.99	0.20	0.22
H	Et	10.05	5.35	0.91	1.10	−0.19	−0.23
Br	Cl	10.14	5.16	1.77	1.96	−0.19	−0.20
Br	Br	10.18	5.25	1.51	1.90	−0.39	−0.41
Cl	Br	10.09	5.21	1.24	1.63	−0.30	−0.41
Cl	Me	10.11	5.26	1.96	1.60	0.36	0.38
Br	Me	10.20	5.30	1.96	1.88	0.08	0.09
Me	Br	10.22	5.27	2.26	2.06	0.20	0.21
Et	F	10.28	4.99	2.82	2.98	−0.16	−0.26
F	H	9.74	5.02	0.44	0.60	−0.16	−0.21

ring in Scheme VIII. These two indices are not correlated; thus, the information from each is a separate statement of the roles of these two structural features. The interpretation of Eq. (7.13) is that the two areas of the molecule that are important for binding are the ring hydroxyl group and the ring methoxyl group oxygen atom. The presence of the OH and OCH_3 groups is constant within the data series but the E-State indices vary in a manner which reflects the effects of the substituents. The equation indicates that as the E-State value of the OH group increases, the activity also increases. In contrast, as the E-State value of the methoxyl oxygen increases, the activity decreases. These two positions are influenced by the substituents at the 3 and 5 positions of the ring. This model does not suggest any direct influence of these substituents on the ipso atoms. The 3 and 5 substituents are not directly involved in the receptor inter-action, but they do influence the interactions of the hydroxyl and methoxyl groups.

An examination of the regression equation, when the variables have been converted to standardized variables, indicates that the relative importance of the 3 substituent to the 5 substituent is about four to one. Furthermore, groups in the 3 position have a stronger effect on the hydroxyl group at position 7 than do groups at the 5 position. These two fragments are clearly capable of inter-action with nearby atoms to form intramolecular hydrogen bonds as is sug-gested by X-ray diffraction analysis (Hogberg et al., 1987).

The model indicates that molecules which could bind to a greater degree might have substituents that will increase the E-State at position 7 and decrease it at position 8. Substituents at the 3 position with small intrinsic values would increase $S(7)$, whereas substituents at position 5 with large intrinsic values would decrease $S(8)$. The application of E-State indexes to the modeling of the binding of a set of salicylamides leads to the development of a statistically strong model for the experimental measure of binding. The E-State indices for two skeletal positions are the basis for the QSAR equation. The two indices, for the ring hydroxyl group and the methoxyl oxygen atom, produce an equation model that explains 89% of the biological activity variance. Furthermore, it is possible to predict structures of candidate molecules which may have enhanced binding by examining variation in the E-State indexes brought about by varia-tion in the two positions of substitution.

IX. BINDING OF CORTICOSTEROIDS

The binding of steroids is an important aspect of their biological behavior. A solid-phase competitive binding assay has been used to measure the affinity constant, K_a, for the binding of a particular set of steroids (Scheme IX) to corticosteroid-binding globulin (CBG) in human plasma at 37°C and pH 7.4

(Dunn *et al.*, 1981; Mickelson *et al.*, 1981; Westphal, 1986). This structure information and activity is expressed as the pK_a in Tables 7.9 and 7.10.

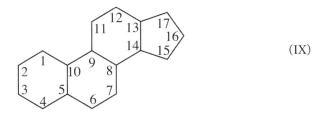

(IX)

In a recent study, a series of 31 steroids were considered where binding affinity varied for CBG (Mickelson *et al.*, 1981). These data have been studied by several other investigators. Cramer *et al.* (1988) used these data in their development of the comparative molecular field analysis (CoMFA). Good *et al.* (1993a,b) obtained results using data matrices derived from similarity based on different indices which gave results comparable to those of Cramer *et al.* Polanski (1997) used molecular similarity indices derived from silhouettes and self-organizing maps to investigate these steroids. Wagner *et al.* (1995) used a three-dimensional autocorrelation descriptor and performed the analysis using neural networks. Kellogg *et al.* (1996) published a pioneering work involving the use of the E-State and a new descriptor, the hydrogen E-State, as a basis for molecular fields in a modified CoMFA study.

DeGregorio (1998) studied these molecules by calculating E-State values for each atom in the molecular scaffold. One structure is different from the rest in an important way. The aldosterone molecule has an uncertain structure since some fraction exists as a ketone while it is usually represented as a hemiacetal. Because of this structure uncertainty, aldosterone was dropped from the list and only 30 were used in this study. The E-State indices were calculated for each atom in the set and were regressed individually and in multiples against the pK_a values for the 30 steroids.

QSAR models were selected for consideration as the best one-, two-, and three-variable equations by examining the E-State indices for the 17 structurally common atoms in the steroid nucleus (Scheme IX). The best three-variable model is

$$\text{p}K_a = 0.318 \ (\pm 0.068) \ S(\text{C-16}) + 1.615 \ (\pm 0.210) \ S(\text{C-4}) \tag{7.14}$$

$$- \ 1.600 \ (\pm 0.197) \ S(\text{C-10}) + 4.077 \ (\pm 0.357)$$

$$r^2 = 0.82, \ s = 0.492, \ n = 30, \ F = 40$$

The three E-State values in this equation are not significantly intercorrelated.

The analysis reveals the importance of position 16 in the steroid nucleus. Although a carbon atom occupies this position in all molecules in the data set,

TABLE 7.9   Corticosteroid-Binding Globulin Data for a Two-Variable Model
with E-State Indices[a]

No.	Structure	X_1	X_2	X_3	X_4	X_5	X_6	X_7	X_8	X_9	X_{10}
1	SB	OH	H	H[b]	H	OH	H				
2	SE	OH	OH	H							
3	SC	=O	H	=O				H	H	H	H
4	SB	H	OH	H[b]	H	=O					
5	SC	=O	OH	COCH$_2$OH	H			H	H	H	H
6	SC	=O	OH	COCH$_2$OH	OH			H	H	H	H
7	SC	=O	=O	COCH$_2$OH	OH			H	H	H	H
8	SE	OH	=O								
9	SC	=O	H	COCH$_2$OH	H			H	H	H	H

its computed E-State value varies with the substitution changes across the
whole molecule.

DeGregorio (1998) observed that coefficients for $S(C\text{-}4)$ and $S(C\text{-}10)$ are
nearly identical but of opposite sign; however, they are not significantly intercorrelated. We may consider that these indices reflect two parts of a structural
singularity, a situation which is sometimes encountered in a classical Free-
Wilson *de novo* analysis. This finding suggests that they relate to a more general
feature of the molecule which is influencing the behavior of the molecule.
A composite variable was included in the analysis of the relationship, with
pK_a as the difference of the two variables $S(C\text{-}4)$ and $S(C\text{-}10)$: $S(C\text{-}4,10) =
S(C\text{-}4) - S(C\text{-}10)$. Using the composite parameter $S(C\text{-}4,10)$, a regression
analysis against the binding affinity (pK_a) produced the best two-variable
equation:

TABLE 7.9 (Continued)

No.	Structure	X_1	X_2	X_3	X_4	X_5	X_6	X_7	X_8	X_9	X_{10}
10	SC	=O	H	COCH$_2$OH	OH			H	H	H	H
11	SB	=O		H[b]	H	OH	H				
12	SD	OH	OH	H	H						
13	SD	OH	OH	H	OH						
14	SD	OH	=O		H						
15	SB	H	OH	H[c]	H	=O					
16	SE	OH	COMe	H							
17	SE	OH	COMe	OH							
18	SC	=O	H	COMe	H			H	H	H	H
19	SC	=O	H	COMe	OH			H	H	H	H
20	SC	=O[d]	H	OH	H			H	H	H	H
21	SF	=O	OH	COCH$_2$OH	OH						
22	SC	=O	OH	COCH$_2$OCOMe	OH			H	H	H	H
23	SC	=O	=O	COMe	H				H	H	H
24	SC	=O	H	COCH$_2$OH	H			OH	H	H	H
25	SC[e]	=O	H	OH	H			H	H	H	H
26	SC	=O	H	COMe	OH			H	OH	H	H
27	SC	=O	H	COMe	H			H	Me	H	H
28	SC[e]	=O	H	COMe	H			H	H	H	H
29	SC	=O	OH	COCH$_2$OH	OH			H	H	Me	H
30	SC	=O	OH	COCH$_2$OH	OH			H	H	Me	F

[a] Structures are drawn according to Kier and Hall (1990), Iizuka et al. (1981), and Burnett (1998).
[b] Of the 5-α steroid series.
[c] Of the 5-β steroid series.
[d] Assumed to be =O (testosterone) as indicated by Iizuka et al. (1981) and not —OH in the table and other publications [see Iizuka et al. (1981) for mistakes in previous publications].
[e] H instead of Me at C$_{10}$ steroid skeleton.

$$pK_a = 0.319 \ (\pm 0.065) \ S(16) + 1.607 \ (\pm 0.145) \ S(4,10) \qquad (7.15)$$
$$+ \ 4.051 \ (\pm 0.228)$$

$$r^2 = 0.821, \ s = 0.483, \ n = 30, \ F = 62$$

$$r^2_{press} = 0.792, \ s_{press} = 0.521$$

The predicted values from this equation are shown in Table 7.10. These QSAR models reveal important aspects of the behavior of these steroid molecules. The equation model reveals two regions in the molecules that are strong contributors to the binding of these steroids at the CBG site. Specifically, the information implicates the A and the D rings of the steroids.

TABLE 7.10 Corticosteroids Binding to Globulin and Two-Variable E-State Model

Observed	Structure [a]	pK_a	Calculated [b]	Residual [c]	Predicted residual [d]
1	SB	5.000	5.31	−0.31	−0.33
2	SE	5.000	5.30	−0.30	−0.33
3	SC	5.763	7.08	−1.32	−1.39
4	SB	5.613	5.26	0.35	0.38
5	SC	7.881	7.42	0.46	0.50
6	SC	7.881	7.34	0.54	0.58
7	SC	6.892	7.42	−0.53	−0.57
8	SE	5.000	5.26	−0.26	−0.28
9	SC	7.653	7.17	0.48	0.51
10	SC	7.881	7.10	0.78	0.83
11	SB	5.919	5.09	0.83	0.92
12	SD	5.000	5.15	−0.15	−0.17
13	SD	5.000	4.72	0.28	0.33
14	SD	5.000	5.11	−0.11	−0.12
15	SB	5.225	5.26	−0.04	−0.04
16	SE	5.225	5.34	−0.12	−0.13
17	SE	5.000	5.26	−0.26	−0.28
18	SC	7.380	7.19	0.19	0.21
19	SC	7.740	7.11	0.63	0.67
20	SC	6.724	7.12	−0.40	−0.42
21	SF	7.512	7.48	0.03	0.03
22	SC	7.553	7.36	0.19	0.20
23	SC	6.779	7.51	−0.73	−0.79
24	SC	7.200	7.42	−0.22	−0.23
25	SC	6.144	6.44	−0.29	−0.31
26	SC	6.247	6.67	−0.43	−0.46
27	SC	7.120	7.02	0.10	0.10
28	SC	6.817	6.50	0.32	0.34
29	SC	7.688	7.41	0.28	−0.29
30	SC	5.797	5.82	−0.03	−0.29

[a] See Table 7.1.

[b] See text for two-variable equation.

[c] Residual = Observed − calculated.

[d] Predicted residual based on the leave-one-out method.

The structure variations in the D ring among the steroids occur as substituents on the 16 or 17 positions. The 17- substituents consist of hydroxyl, keto, or acyl groups. The 16 position is occasionally substituted with a methyl or a hydroxyl group. Thus, the E-State index for the 16 position encodes information about the substitution pattern in the D ring. Lower values for $S(C\text{-}16)$ correspond to lower binding affinity. This trend of E-State values at position 16 is a consequence of more electronegativity at or near positions 16 and 17. For example, the presence of an acyl group at the 17 position leads to greater binding (and a higher E-State) than that of a hydroxyl group at position 17. These relationships between substitution patterns on the D ring and binding signaled by the 16 position E-State are evidence of a pharmacophore feature associated with the D ring.

The second area implicated by the QSAR model is the A ring. The second variable in the equation model indicates the relationship of the binding affinity to an index formed from the difference between $S(C-4)$ and $S(C-10)$. This is another example of a composite index that has been used in other studies employing the E-State (Hall *et al.*, 1991a,b). This index, labeled $S(C-4,10)$, carries information about the electronic and substitution pattern at the 4 and 10 positions and also, perhaps more important, at all positions in the A ring. Like the example described for $S(16)$ signaling the effects on the D ring, the $S(C-4,10)$ composite E-State index signals structural changes on the A ring governing binding affinity. This means that a pharmacophore feature resides in the A ring as well as in the D ring.

The study reinforces an important aspect of the E-State identification of structure features. A region of a molecule which may be part of a pharmacophore may be illuminated by a composite E-State index made up of the difference (or the sum) of two E-State indices.

X. EXERCISES

To complete this chapter, the reader is invited to participate in the QSAR study of data taken from the literature but not as yet analyzed. These exercises provide additional experience with the compact disc included in the book and with the techniques of data entry, running a program, harvesting the indices, and correlating these with the biological or physical data. The reader should be able to find a way to carry out the regression analysis.

Case 1: Anesthetic Haloalkanes

A study by Larsen (1969) reported the minimum anesthetic concentrations (Table 7.11) of a series of trifluroethyl halides (Scheme X). The activities, expressed as log MAC, cover a range of two log units and therefore are suitable for an analysis of this type. The goal is to find an atom E-State index for an atom that changes in parallel with the potency within the series. This may be followed by an interpretation and some predictions of structures that may have high potency.

TABLE 7.11   Anesthetic Concentration of
Trifluoromethyl Ethanes

Y	Z	S(F)	log MAC [a]
H	H	10.410	1.60
H	Cl	10.653	0.90
H	Br	10.738	0.45
H	I	10.777	0.10
F	F	10.428	1.70
F	Cl	10.671	1.18
F	Br	10.756	0.70
F	I	10.796	0.30
Cl	Cl	10.914	0.43
Cl	Br	11.000	-0.10
Br	Br	11.084	-0.40

[a] Log of the minimum anesthetic concentration.

(X)

Case 2: Benzoxazinone Elastase Inhibitors

A series of benzoxazinones (Scheme XI) have been reported to be active as elastase enzyme inhibitors (Krantz et al., 1987). The structures and inhibition constants, log K_i, are shown in Table 7.12. The reader is challenged to deter-

TABLE 7.12   Benzoxazinone Elastase Inhibitors

R_1	R_2	log K_i
H	H	1.80
CH$_3$	H	0.84
C$_2$H$_5$	H	-1.29
C$_2$H$_5$	CH$_3$	-1.47
C$_3$H$_7$	H	-1.04
C$_3$H$_7$	CH$_3$	-2.22
i-C$_3$H$_7$	H	-1.28
i-C$_3$H$_7$	CH$_3$	-2.82
i-C$_3$H$_7$	C$_2$H$_5$	-3.03
C$_4$H$_9$	H	-1.08
C$_4$H$_9$	CH$_3$	-1.64
C$_4$H$_9$	C$_2$H$_5$	-2.38
sec-C$_4$H$_9$	H	-1.37

mine which atom E-State value among the four heteroatoms best relates to the inhibitory potency. Is it possible to design a molecule that may be more potent from this information?

(XI)

XI. SUMMARY

In this chapter, several examples have been given which reveal the direct use of the atom-level E-State indices in the development of strong QSAR models. The topological superposition approach indicates which atom E-State indices for a set of molecules directly parallel the measured property. In this manner, one can suggest which atoms in a molecule are most important for determining property values. This finding has a direct bearing on the definition of the phamacophore for that activity. Furthermore, because the structure influences of substituents are readily apparent in the E-State formalism, it is straightforward to identify structure changes which may enhance the property value. Atom-level E-State indices provide a clear picture of important regions in the structure and a basis for a modification strategy in drug design.

XII. A LOOK AHEAD

In this chapter, we developed models which use only the atom-level E-State indices. Chapters 8 and 9 present investigations in which other topological indices were also employed. Other indices may represent broader aspects of the molecular structure or more specific features such as the branching pattern, the character of a ring system, or the overall shape of the molecule. Such combinations of indices provide a powerful basis for development of models and insight into the relation between structure and properties.

REFERENCES

Amoore, J. E. (1967). Specific anosmia: A clue to the olfactory code. *Nature* 214, 1095–1098.
Burnett, J. (1998). Imidazole inhibitors of thromboxane synthase: QSAR analysis using the E-State. Personal communication.

Cramer, R. D., Patterson, D. E., and Bunce, J. D. (1988). Comparative molecular field analysis (CoMFA). Effect of shape on binding of steroids to carrier proteins. *J. Am. Chem. Soc.* **110**, 5959–5967.

DeGregorio, C., Kier, L. B., and Hall, L. H. (1998). QSAR modeling with the electrotopological state indices: Corticosteroids. *J. Comp.-Aided Mol. Des.* **12**, 557–561.

Dunn, J. F., Nisula, B. C., and Rodbart, D. (1981). Transport of steroid hormones: Binding of 21 endogenous steroids to the testosterone-binding globulin and corticosteroid-binding globulin in human plasma. *J. Clin. Endocrinol. Metab.* **53**, 58–68.

Elguero, J., Marzin, C., Katritsky, A. R., and Linda, P. (1976). Purines containing two potential tautomeric groups. *Adv. Hetero. Chem.* **S1**, 519–525.

Fulcrand, P., Berge, G., Noel, A.-M., Chevallet, P., Castel, J., and Orzalesi, H. (1978). Hydrazide inhibitors of monoamine oxidase: Correlations with Hansch and free Wilson. *Eur. J. Med. Chem.* **13**, 177–182.

Good, A. C., Peterson, S. J., and Richards, W. G. (1993a). QSARs from similarity matrices. Technique validation and application in the comparison of different similarity evaluation methods. *J. Med. Chem.* **36**, 2929–2937.

Good, A. C., So, S. S., and Richards, W. G. (1993b). Structure–activity relationships from molecular similarity matrices. *J. Med. Chem.* **36**, 433–438.

Hall, L. H., and Kier, L. B. (1978). Molecular connectivity and substructure analysis. *J. Pharm. Sci.* **67**, 1743–1749.

Hall, L. H., and Kier, L. B. (1992). Binding of salicylamides: QSAR analysis with E-State indexes. *Med. Chem. Res.* **2**, 497–501.

Hall, L. H., Mohney, B. K., and Kier, L. B. (1991a). The electrotopological state: Structure information at the atomic level for molecular graphs. *J. Chem. Inf. Comput. Sci.* **31**, 76–83.

Hall, L. H., Mohney, B. K., and Kier, L. B. (1991b). The electrotopological state: An atomic index for QSAR. *Quant. Structure–Activity Relat.* **10**, 43–48.

Hogberg, T., Norinder, U., Ramsby, S., and Stensland, B. (1987). Crystallographic, theoretical and molecular modeling studies on the conformations of the salicylamide, raclopride, a selective dopamine-D2 antagonist. *J. Pharm. Pharmacol.* **39**, 787–796.

Iizuka, K., Akahane, K., Momose, D., and Nakazawa, M. (1981). Highly selective inhibitors of thromboxane synthetase. 1. Imidazole derivatives. *J. Med. Chem.* **24**, 1139–1148.

Jacobsen, K. A., Kiriasis, L., Barone, S., Bradbury, B. J., Vdai, K., Campagne, J. M., Secunda, S., Daly, J. W., Neumeyer, J. L., and Pfleiderer, W. (1989). Sulfur-containing 1,3-dialkylxanthine derivatives as selective antagonists at A_1-adenosine receptors. *Med. Chem.* **32**, 1873–1879.

Joshi, N., and Kier, L. B. (1992). An E-State analysis of adenosine A1 inhibitors. *Med. Chem. Res.* **1**, 409–412.

Kellogg, G. E., Kier, L. B., Guillard, P., and Hall, L. H. (1996). A-State fields: Applications to 3-D QSAR. *J. Comp.-Aided Mol. Design* **10**, 513–520.

Kier, L. B. (1986). Shape indexes of orders one and three from molecular graphs. *Quant. Structure–Activity Relat.* **5**, 1–8.

Kier, L. B., and Hall, L. H. (1990). An electrotopological state index for atoms in molecules. *Pharm. Res.* **7**, 801–809.

Krantz, *et al.* (1987). Design of alternative substrate inhibitors of serine proteases. Synergistic use of alkyl substitution to impede enzyme-catalysed deacylation. *J. Med. Chem.* **30**, 591–598.

Larsen, E. R. (1969). Fluorine compounds in anesthesiology. *Fluorine Chem. Rev.* **3**, 43–48.

Lopata, A., Darvas, F., Stadler-Szoke, A., and Szejtli, J. (1985). Quantitative structure stability relationships among inclusion complexes of cyclodextrins. I: Barbituric acid derivatives. *J. Pharm. Sci.* **74**, 211–213.

Mickelson, K. E., Forsthoefel, J., and Westphal, U. (1981). Steroid–protein interactions. Human corticosteroid binding globulin: Some physicochemical properties and binding specificity. *Biochemistry* **20**, 6211–6218.

Ogren, S.-O., Hall, H., Kohler, C., Magnusson, O., and Sjorstand, S. E. (1986). The selective dopamine D2 receptor antagonist, raclopride, discriminates between dopamine mediated motor functions. *Psychopharmacology* **90**, 287–294.

Peet, N., Lentz, N. L., Meng, E. C., Dudley, M. W., Ogden, A. M. L., Demeter, D. A., Weintraub, H. J. R., and Bez, P. (1990). A novel synthesis of xanthines: Support for a new binding mode for xanthines with respect to adenosine at adenosine receptors. *J. Med. Chem.* **33**, 3127–3130.

Polanski, J. J. (1997). The receptor-like neural network for modeling corticosteroids and testosterone binding globulins. *J. Chem. Inf. Comput. Sci.* **37**, 553–561.

Stone, T. W. (1989). Purine receptors and their pharmacological roles. *Adv. Drug Res.* **18**, 292–429.

Tamm, I., Folkers, K., Shunk, C., and Horofall, F. L. (1953). Inhibition of influenza virus multiplication by alkyl derivatives of benzimidazoles. *J. Exp. Med.* **98**, 219–229.

Teranishi, R., Buttery, R. G., and Guadagni, D. G. (1974). Odor quality and chemical structure in fruit and vegetable odors. *Ann. N. Y. Acad. Sci.* **237**, 209–216.

Tsantili-Kakoulidou, A., and Kier, L. B. (1992). A QSAR study of alkylpyrazine odor modalities. *Pharm. Res.* **9**, 1321–1324.

Uekama, K., Hirayama, F., Nasu, S., Matsuo, N., and Irie, T. (1978). Determination of the stability constants for inclusion complex of cyclodextrins with various drug molecules by high performance liquid chromatography. *Chem. Pharm. Bull.* **26**, 3477–3484.

Van Galen, P. J. M., Van Vljimen, H. W. T., Ljerman, A. P., and Soudjin, W. (1990). A model for the antagonist binding site on the adenosine A1 receptor, based on steric, electrostatic and hydrophobic properties. *J. Med. Chem.* **33**, 1708–1713.

Wagner, M., Sadowski, J., and Gasteiger, J. (1995). Autocorrelation of molecular surface properties for modeling corticosteroid binding globulin and cytosolic receptor activity by neural networks. *J. Am. Chem. Soc.* **117**, 7769–7778.

Westphal, U. (1986). *Steroid–Protein Interactions II*. Springer-Verlag, Berlin.

CHAPTER 8

Structure–Activity Studies: Mixed Indices

I. ANALYSES WITH A MIXED SET OF INDICES

A common process in structure–activity analysis is the blending of different categories of structure descriptors to fully express structure variation as well as to achieve a quality model. This is the basis of the physical property-based Hansch approach (Hansch, 1969). In this scheme, the functions of a drug molecule are modeled as an ensemble of effects summarized as electronic, steric, and hydrophobic. The structure–activity quantitation of this classical model is built around quantifiable properties reflecting each of these phenomena. It is clear that these are not orthogonal effects. The steric structure and its influence on the molecular properties are the result of electronic structure manifesting itself through repulsive and attractive forces. The hydrophobic behavior of a molecule is a consequence of both electronic and steric attributes. It is convenient for the chemist to parse molecular behavior into these categories, which can then be examined in a reductionist mode.

The Hansch approach employs physical properties as surrogates of structure for these three categories in the general model. The electronic influence on chemical behavior is often quantified using the Hammett substituent constant derived from the reaction equilibria measured for a reference molecule and variants on this structure (Hammett, 1940). The partition coefficient is recorded as the ratio of solute concentrations in two immiscible liquids when a thorough mixing among the ingredient has preceded the equilibrium state. The steric effect may be modeled from the relative reaction rates of some process in which the inhibiting influence of a group is quantified (Taft, 1956).

A combination of these three property parameters has been frequently used in structure–activity equations in the search for a model. These models have achieved success and have been the bases of molecular design in a number of technology areas. Where the property composing the substituent index or hydrophobic effect has not been available, predicting schemes have been employed based on fragment estimates of their contribution to a property. The achievement of a successful structure–activity model by itself is not sufficient to proceed directly to molecular design. The indices encoding the physical properties must be converted to structure information in order to begin the design process. This conversion is not always easy and represents a deficiency in this general approach. This problem is not present in structure–activity models built around structure-based indices, such as the molecular connectivity (Kier and Hall, 1976a, 1986), kappa shape (Kier, 1987), and the E-State indices. At times it has proven useful to combine these structure indices in the search for good structure representation. In this chapter, we describe the use of some mixed-index equations and their interpretation.

The use of a mixed set of parameters characterizing molecular structure was introduced by us in a series of papers developing the molecular connectivity indices (Kier *et al.*, 1976; Kier and Hall, 1976b). The premise of the molecular connectivity paradigm was that the approach to meaningful encoding of structure is through the use of nearly independent indices derived from the non-empirical weighted counts of paths of various lengths and architecture. The families of indices generated by this method lend themselves ideally to the creation of a profile of structure which may relate to a property or activity and which can be inverted to information about subsequent molecular design. Numerous structure–activity studies have been reported using ensembles of these indices (Richard and Kier, 1980; Hall and Kier, 1984, 1986).

Among these multivariable analyses are a number of studies using E-State indices as one participant. These are usually comingled with molecular connectivity topological indices or kappa shape indices. The general interpretation has been based on the presumption of a role for the whole molecule along with a role for a specific atom or group. This idea is consistent with common use of a pharmacophore and a molecular hydrophobic character to describe the essential features of a molecule invoking some activity. In this chapter, several of these multivariate studies will be examined as models for general applications. In keeping with the theme of the book, emphasis is on the E-State and its interpretation. However, the context of the E-State indices is within the molecular description portrayed by the other indices and so interpretation must be integrated with the entire equation.

II. REACTION OF THE OH RADICAL WITH CHLOROFLUOROCARBONS

A good example of the use of the E-State method in combination with the molecular connectivity and the kappa shape methods was published by Kier and Hall (1995). In this study, a series of chlorofluorocarbons (CFCs; Scheme I) was studied for the effect of their reaction with the hydroxyl radical.

$$(C_xCl_yF_x)_n \tag{I}$$

A regression analysis revealed that the best three-variable equation included the $^3\chi_c$ molecular connectivity index, the $^1\kappa_\alpha$ shape index, and the E-State index of the carbon atom bonding the largest number of hydrogen atoms, $S(1)$. The relating equation was determined to be

$$\log k = 0.377\ S(1) - 1.346\ ^3\chi_c + 0.671\ ^1\kappa_\alpha + 5.415 \qquad (8.1)$$

$$r^2 = 0.89,\ s = 0.27,\ F = 60,\ n = 26$$

The three indices are not significantly intercorrelated. The data with predicted values are shown in Table 8.1. The equation model was developed from three variables predicting almost 90% of the variation in the rates of reaction. The three parameters are representative of three attribute classifications described by nonempirical structure indices. The molecular connectivity index, $^3\chi_c$, encodes the extent of skeletal branching. In Eq. (8.1), the information that

TABLE 8.1   Rates of Hydrogen Abstraction in CFCs

Molecule	$S(1)$	$^3\chi_c$	$^1\kappa_\alpha$	$\log K$	Calculated $\log K$
CH_3Cl	1.472	0.000	2.29	7.36	7.50
CH_2Cl_2	0.194	0.000	3.58	8.00	7.88
$CHCl_3$	−0.750	0.577	4.87	7.80	7.63
CH_3F	0.500	0.000	1.93	6.95	6.90
CH_2F_2	−1.750	0.000	2.86	6.81	6.67
CHF_3	−3.677	0.577	3.79	5.10	5.79
CH_2ClF	−0.788	0.000	3.22	7.46	7.29
$CHCl_2F$	−1.722	0.577	4.51	7.30	7.03
$CHClF_2$	−2.694	0.577	4.15	6.45	6.40
CH_3CH_2Cl	1.890	0.000	3.29	8.37	8.34
CH_3CH_2F	1.458	0.000	2.93	8.14	7.93
CH_3CHCl_2	1.698	0.577	4.58	8.20	8.34
CH_3CHF_2	0.833	0.577	3.86	7.48	7.54
CH_2FCH_2F	−0.847	0.000	3.86	7.83	7.69
$CH_2ClCHCl_2$	−0.531	0.408	5.74	8.28	8.51
CH_2FCHF_2	−1.528	0.408	4.79	7.47	7.50
CH_2ClCF_2Cl	−0.825	1.560	6.44	7.20	7.32
CH_2ClCF_3	−1.257	1.560	6.08	6.95	6.92
CH_2FCF_3	−1.313	1.560	5.59	6.70	6.57
CHF_2CHF_2	−3.418	0.666	5.72	6.50	7.04
$CHCl_2CF_3$	−2.257	1.654	7.37	7.40	7.29
$CHClFCF_3$	−3.229	1.654	7.01	6.87	6.68
CHF_2CF_3	−4.201	1.654	6.65	6.48	6.07
CH_3CCl_3	1.484	2.000	5.87	6.80	7.21
CH_3CF_3	0.187	2.000	4.79	5.95	6.00
CH_3CF_2Cl	0.620	2.000	5.15	6.60	6.42

branching in the CFC is detrimental to the OH radical attack at the abstraction site, probably due to steric crowding, is encoded. The information encoded in the $S(1)$ term in the equation reveals that the influence imparted by highly electronegative atoms in the vicinity of the abstraction site is detrimental to the reaction with the OH radical. This is a consequence of the polarization of nearby C—H bonds, an event inimical to the radical scission of this bond by the OH radical. The kappa index encodes the information that higher molecular weight halogens facilitate the abstraction reaction. Chlorine promotes a faster reaction than fluorine since its lower electronegativity supports the free radical scission of the CH bond better than the influence of a nearby fluorine atom. This effect may be due to the possibility that an electronegative influence on the C—H bond is facilitating a scission leading to H^+ rather than $H\cdot$.

The model based on a chi, a kappa, and an E-State index is statistically sound and may be sufficiently robust for predictive use. The interpretation of the equation is straightforward, giving some insight into the mechanism of the radical attack. The model has a significant potential for the prediction of this important reaction.

III. 5-HT$_2$ BINDING OF PHENYLISOPROPYLAMINES

Grella (1998) examined the structure–activity relationships of several phenylisopropylamines (Scheme II) that bind with the 5-HT$_2$ receptor.

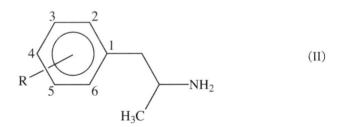

(II)

A series of these compounds, substituted in the phenyl ring, has been reported by Arvidsson *et al.* (1986). A two-variable equation was found relating the log K_i, the binding affinity, to the E-State index for a specific carbon atom and a molecular connectivity index:

$$\log K_i = 0.674\ S(C\text{-}2) - 3.526\ ^4\chi^v_{pc} + 5.333 \tag{8.2}$$

$$r^2 = 0.86,\ s = 0.433,\ F = 35,\ n = 15$$

TABLE 8.2   5-HT$_2$ Receptor Ligands

						log K_i	
R_2	R_3	R_4	R_5	$^4\chi^v_c$	S(C-2)	Observed	Calculated
H	H	OCH$_3$	H	0.518	2.127	4.53	4.94
OCH$_3$	H	H	OCH$_3$	0.721	0.947	3.72	3.43
H	OCH$_3$	OCH$_3$	H	0.663	2.044	4.64	4.37
OCH$_3$	H	OCH$_3$	OCH$_3$	0.866	0.856	3.22	2.86
H	OCH$_3$	OCH$_3$	OCH$_3$	0.795	2.005	4.22	3.88
OCH$_3$	H	OC$_2$H$_5$	OCH$_3$	0.866	0.856	3.35	2.86
OCH$_3$	H	CH$_3$	OCH$_3$	0.969	0.964	2.00	2.57
OH	H	CH$_3$	OCH$_3$	0.992	1.089	2.85	2.57
OCH$_3$	H	C$_2$H$_5$	OCH$_3$	0.991	0.981	2.00	2.50
OCH$_3$	H	C$_3$H$_7$	OCH$_3$	0.961	0.992	1.84	2.61
OCH$_3$	H	C$_4$H$_9$	OCH$_3$	0.961	0.999	2.32	2.62
OCH$_3$	H	F	OCH$_3$	0.799	0.589	3.05	2.91
OCH$_3$	H	Br	OCH$_3$	1.233	0.917	1.80	1.60
OCH$_3$	H	I	OCH$_3$	1.389	0.956	1.28	1.08
OCH$_3$	H	NO$_2$	OCH$_3$	0.911	0.537	2.48	2.48

The two indices are not significantly correlated. The calculated results are shown in Table 8.2. The equation accounts for 86% of the variation in the experimental log K_i values. The $^4\chi^v_{pc}$ encodes the degree and the location of branching on the phenyl ring (Kier, 1980). The model reveals that binding tends to increase with increasing values of this index. This increase can be interpreted in terms of structure to indicate that more substituents and more closely positioned phenyl substituents lead to lower binding. The E-State index for the No. 2 carbon in Scheme II is most sensitive to the influence of nearby electronegative atoms or groups. Lower S(C-2) values lead to lower K_i values reflecting the roles of electronegative substituents near the 2 position.

This study demonstrates the value of the E-State index in identifying a specific atom implicated in an activity. It also illustrates the importance of the molecular-level structure influencing a property such as binding.

IV. OPIATE RECEPTOR ACTIVITY OF FENTANYL-LIKE COMPOUNDS

A series of fentanyl-like compounds (Scheme III) have been tested for their opiate activity (Burnett, 1998). Burnett carried out a study of a subset of this series in which a methyl ester group was a common feature (van Daele et al., 1995).

(III)

A number of structure indices, including E-State and molecular connectivity indices, were used in a structure–activity analysis. The results of this study produced the following equation model:

$$\log 1/ED_{50} = 7.408\ S(C\text{-}14) - 0.740\ ^3\chi_c - 86.510 \qquad (8.3)$$

$$r^2 = 0.82,\ s = 0.452,\ F = 30,\ n = 16$$

The predicted values are shown in Table 8.3. There is no significant correlation between the variables. The molecular connectivity $^3\chi_c$ index describes the ex-

TABLE 8.3   Opioid Activity of Fentanyl-like Compounds

			-log ED_{50}	
R	$^3\chi_c$	S(C-14)	Observed	Calculated
CH_3	1.516	12.596	5.48	5.68
C_2H_5	1.431	12.684	5.83	6.43
C_3H_7	1.431	12.754	7.32	6.92
i-C_3H_7	1.727	12.766	6.94	6.74
C_4H_9	1.431	12.810	7.56	7.36
C_5H_{11}	1.431	12.856	8.16	7.66
C_6H_{13}	1.431	12.895	8.19	7.99
C_7H_{15}	1.431	12.928	7.77	8.17
$C_6H_5-CH_2CH_2$	1.635	12.975	9.07	8.37
$C_6H_5-CH(OH)CH_2$	1.834	12.990	8.70	8.40
$C_6H_5-CH_2CH(CH_3)$	1.834	13.057	8.82	8.82
$C_6H_5-CH_2CH_2CH_2$	1.635	12.995	7.78	8.58
4-Cl—$C_6H_5-CH_2CH_2$	1.924	12.995	8.02	8.32
4-$CH_3-C_6H_5-CH_2CH_2$	1.924	13.007	7.95	8.45
3-$CH_3-C_6H_5-CH_2CH_2$	1.924	13.013	8.40	8.50
2-$CH_3-C_6H_5-CH_2CH_2$	2.358	13.031	8.58	8.28

tent of skeletal branching in the R group—generally increasing with increased degree of branching and with more branch points in the skeleton. Represented in this fashion, branching is detrimental to the activity of the compound, as reflected in the negative coefficient of the index. The information encoded in the $S(C-14)$ term reveals the influence imparted by the R substituents. The modulation of the relative polarity of the ester group, contributed by the R substituents, is a critical factor in the magnitude of the effective dose for this series of molecules.

V. E-STATE AND BIMOLECULAR ENCOUNTER PARAMETERS

Most physical properties arise from intermolecular encounters of molecules, governed by the pattern of atoms and groups on the mantle of each. We considered these molecular surface encounters and formulated a series of parameters encoding the pattern information. One model of these encounters may be created by considering the encounter of just two molecules (butane is used as an example; Scheme IV).

$$H_3C\!\!-\!\!CH_2\!\!-\!\!CH_2\!\!-\!\!CH_3 \qquad\qquad (IV)$$

The encounter of two molecules of butane is a function of the mantle hydrogen atoms on each, considered either singly or in multiples. A series of bimolecular encounter parameters, discerned from Scheme IV, are useful in describing a structure index of possible value.

The number of possible H-to-H encounters between any two molecules in Scheme IV is 100. This particular interaction is coded as H_H for convenience. A more information-rich parameter is the count of hydrogen atoms encountering geminal pairs of hydrogens on the other molecule, H_H_2. In Scheme IV, there are 160 of these encounters possible. In contrast, there are 40 encounters involving a hydrogen and three geminal hydrogens on the other molecule, H_H_3. These bimolecular encounter parameters carry significant information about the peripheral topology of molecules and their encounter accessibility. Some of these parameters are shown in Table 8.4.

In addition to these geometric possibilities, the E-State indices of electron-rich atoms on a molecule must certainly be descriptive of the strength and likelihood of an intermolecular encounter. The coupling of an E-State index with one or more bimolecular encounter parameters is a good strategy to use when studying structure–physical property relationships. Two studies illustrate this approach using the boiling point as a target property.

TABLE 8.4 Possible Bimolecular Encounter pairs for Alkanes

Alkane	H_H	H_H$_2$	H_H$_3$	H$_2$_H$_2$	H$_2$_H$_3$	H$_3$_H$_3$
Methane	16	48	32	36	48	16
Ethane	36	72	24	36	24	4
Propane	64	112	32	49	28	4
Butane	100	160	40	64	32	4
Isobutane	100	180	60	81	54	9
Pentane	144	216	48	81	36	4
Isopentane	144	240	72	100	60	9
Neopentane	144	288	96	144	96	16
Hexane	196	196	56	100	40	4
3-Methyl pentane	196	308	84	121	66	9
2-Methyl pentane	196	308	84	121	66	9
2,3-Dimethyl butane	196	336	112	144	96	16
2,2-Dimethyl butane	196	344	112	169	104	16

A. PRIMARY AMINE BOILING POINTS

There are two aspects of structure within a series of molecules, such as the primary amines, that influence a property such as the boiling point. The first is the electron-rich heteroatom—the nitrogen and its immediate neighbors. The second is the topology and shape of the less interactive part of the molecule. The structure information about each fragment may be encoded using the E-State indices and the bimolecular encounter parameters, respectively. A set of 21 primary amines from earlier work (Kier and Hall, 1976b) were chosen for this analysis. The H_H$_2$, H_H$_3$, and H_N counts were recorded for each compound. In addition, the E-State of the amine group $S(—NH_2)$ was calculated. The equation from regression analysis is

$$\text{bp (°C)} = -0.727 \ (\pm 0.079) \ (\text{H\_H}_2)$$
$$- 0.278 \ (\pm 0.047) \ (\text{H\_H}_3)$$
$$+ 10.716 \ (\pm 0.717) \ (\text{N\_H}) \tag{8.4}$$
$$- 34.096 \ (\pm 9.477) \ S(—NH_2)$$
$$+ 42.066 \ (\pm 38.250)$$

$$r^2 = 0.998, \ s = 3.1, \ F = 1647, \ n = 21$$

The data are shown in Table 8.5. The combination of the E-State and descriptors encoding the topology of the nonpolar parts of the molecules is very useful in modeling the boiling point of these compounds.

TABLE 8.5 Primary Amine Boiling Points and E-State and Bimolecular Encounter Indices

Amine–R	$S(-NH_2)$	bp (°C)	Calculated bp (°C)
CH_3-	4.500	−7.5	−9.0
CH_3CH_2-	4.847	17.0	16.9
$(CH_3)_2CH-$	5.111	34.0	39.2
$CH_3CH(CH_3)CH_2-$	5.171	68.0	70.7
$CH_3CH_2CH(CH_3)-$	5.292	63.0	65.9
$CH_3CH_2CH_2-$	5.028	49.0	46.0
$CH_3CH_2CH_2CH_2-$	5.139	77.8	74.9
$CH_2CH_2C(CH_3)_2-$	5.535	78.0	79.2
$CH_3CH(CH_3)CH_2CH_2-$	5.229	95.0	98.9
$(CH_3CH_2)_2CH-$	5.472	91.0	89.5
$CH_3(CH_2)_4-$	5.215	104.4	102.8
$CH_3C(CH_3)_2CH(CH_3)-$	5.569	102.0	98.6
$CH_3(CH_2)_5-$	5.311	130.0	129.4
$CH_3(CH_2)_4CH(CH_3)-$	5.478	142.0	140.4
$CH_3(CH_2)_6-$	5.311	156.9	153.9
$CH_3(CH_2)_3CH(CH_3)-$	5.478	117.5	115.7
$CH_3CH_2CH(CH_3)CH_2CH(CH_3)-$	5.569	132.5	134.0
$CH_3CH_2CH_2CH(CH_3)-$	5.403	92.0	91.8
$CH_3(CH_2)_7-$	5.343	176.0	176.6
$CH_3(CH_2)_8-$	5.369	192.0	196.8
$CH_3(CH_2)_9-$	5.391	217.0	215.4

B. ALCOHOL BOILING POINTS

A second study using this same approach considered 28 alcohols previously analyzed using topological indices (Kier and Hall, 1976b). The H_H$_2$, H_O, and S(OH) E-State values were determined to be important:

$$bp \ (°C) = 0.0800 \ (\pm0.031) \ (H\_H_2)$$
$$+ \ 4.602 \ (\pm0.49) \ (H\_O)$$
$$- \ 22.413 \ (\pm3.85) \ S(-OH)$$
$$+ \ 198.02 \ (\pm25.29)$$

(8.5)

$$r^2 = 0.926, \ s = 5.8, \ F = 204, \ n = 28$$

TABLE 8.6   Alcohol Boiling Points versus E-State and Bimolecular Encounter
Indices

Alcohol–R	$S(—OH)$	bp (°C)	Calculated bp (°C)
$(CH_3)_2CH—$	8.056	82.4	91.1
$CH_3CH_2CH_2—$	7.875	97.1	97.4
$CH_3(CH_2)_3—$	8.066	117.6	113.6
$CH_3CH(CH_3)CH_2—$	8.144	108.1	109.0
$CH_3CH_2C(CH_3)_2—$	8.826	102.3	112.4
$CH_3CH_2CH_2CH(CH_3)—$	8.552	118.9	120.3
$CH_3CH(CH_3)CH_2CH_2—$	8.237	132.0	127.4
$CH_3CH_2CH(CH_3)CH_2—$	8.335	128.9	125.2
$CH_3(CH_2)_4—$	8.197	138.0	131.8
$CH_3C(CH_3)_2CH(CH_3)—$	9.095	120.4	123.0
$CH_3(CH_2)_2C(CH_3)_2—$	9.018	123.0	128.9
$(CH_3CH_2)_2C(CH_3)—$	9.132	136.0	126.3
$CH_3CH_2C(CH_3)_2CH_2—$	8.594	136.7	138.4
$CH_3CH(CH_3)CH_2CH(CH_3)—$	8.723	133.0	133.4
$CH_3CH(CH_3)CH(CH_3CH_2)—$	8.935	127.5	128.7
$CH_3CH(CH_3)CH(CH_3)CH_2—$	8.505	145.0	138.3
$CH_3CH_2CH_2CH(CH_3)CH_2—$	8.466	148.0	143.4
$CH_3(CH_2)_4—$	8.293	157.5	169.8
$(CH_3CH(CH_3))_2CH—$	9.204	140.0	139.0
$CH_3CH(CH_3)CH_2CH(CH_3)—$	8.584	159.8	157.7
$(CH_3CH_2)_3C—$	9.438	142.0	138.5
$CH_3(CH_2)_6—$	8.366	176.8	172.2
$(CH_3CH_2)_2(CH_3)C—$	9.514	161.0	161.3
$(CH_3(CH_2)_3)(CH_3CH_2)(CH_3)C—$	9.454	163.0	162.7
$CH_3CH(CH_3)CH_2(CH_2)_4—$	8.431	188.0	188.3
$CH_3(CH_2)_7—$	8.423	194.4	193.0
$CH_3(CH_2)_5C(CH_3)_2—$	9.137	178.0	188.4
$(CH_3CH_2CH_2)_2(CH_3CH_2)C—$	9.820	182.0	177.0

This is a satisfactory model, considering that there is a good mix of 14 primary, 6 secondary, and 8 tertiary alcohols. The data are shown in Table 8.6.

VI. DOPAMINE RECEPTOR LIGANDS

Schieck (1998) reported on a structure–activity analysis of dopamine D_3 ligands tested by Stjernlof et al. (1995). Fifteen benzindoles with the general
structure shown in Scheme V were selected from this data set and analyzed
using the E-State and topological indices (Table 8.7).

TABLE 8.7   Dopamine Receptor Ligands

R_1	R_2	R_3	R_4	$^4\chi^v_{pc}$	$S(C\text{-}4)$	$\log K_i$	Calculated $\log K_i$
H	H	H	H	1.56	2.18	1.57	1.05
CHO	H	H	H	1.72	2.07	1.85	1.73
CN	H	H	H	1.69	2.07	1.63	1.62
CH_2CF_3	H	H	H	1.84	1.88	2.13	2.28
H	H	H	F	1.63	−0.16	0.20	0.81
CHO	H	H	F	1.78	−0.30	1.81	1.49
CN	H	H	F	1.76	−0.03	1.80	1.45
C_3H_7	H	H	F	1.86	−0.14	2.05	1.87
H	H	CH_3	F	1.89	−0.07	1.72	2.04
CHO	H	CH_3	F	2.03	−0.21	2.76	2.68
$COCF_3$	H	H	H	1.94	1.73	2.83	2.68
$COCH_3$	H	H	H	1.87	2.08	2.51	2.46
H_2	=O	H	H	1.71	2.09	1.36	1.72
CH_3	H	H	H	1.80	2.21	1.83	2.14
CHO	H	CH_3	H	1.98	2.18	3.01	3.00

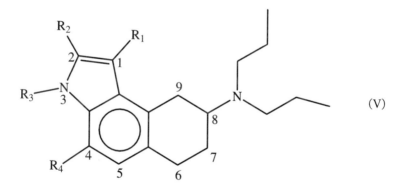

(V)

A two-variable equation was discovered to give a good correlation with the binding affinity, $\log K_i$

$$\log K_i = 4.66\ (\pm 0.68)\ ^4\chi^v_{pc} + 0.235$$
$$(\pm 0.079)\ S(C\text{-}4) - 6.75\ (\pm 1.26)$$

(8.6)

$$r^2 = 0.81,\ s = 0.11,\ F = 25,\ n = 15$$

The predicted values from Eq. (8.6) are shown in Table 8.7. The $^4\chi^v_{pc}$ index, noted earlier in the chapter, encodes information about the number and proximity of ring substituents (Kier, 1980). From the equation model, it can be seen that the log K_i value is significantly influenced by the overall topology and the unsubstituted benzene ring position in Scheme V. Electronegative groups on the molecule near the benzene ring produce lower $S(C\text{-}4)$ values, which correlates with a better binding. The ring 4 position, implicated in the equation model, cannot necessarily be identified as a binding site. It is a structural feature that responds to the changing structures in the series in such a way that it parallels the measured activity. This "signal" is of great value in predicting candidate molecules for further study.

VII. CARBOQUINONES AS ANTILEUKEMIC AGENTS

A set of 34 carboquinones (Scheme VI) taken from Ickikawa and Aoyama (1991) were analyzed by Gough and Hall (1998a) using the combination of molecular connectivity chi, kappa shape, and E-State indices.

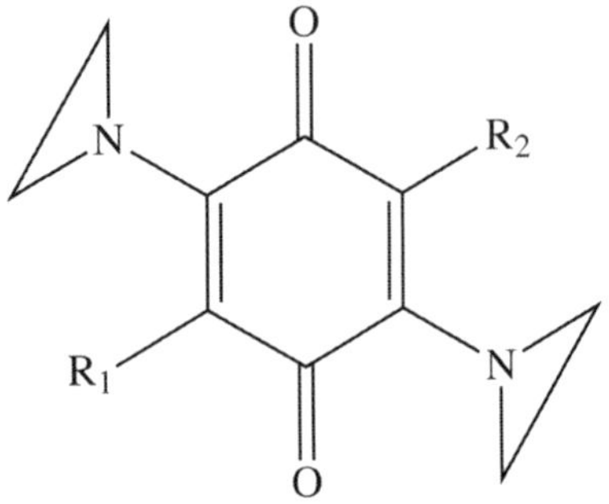

(VI)

A four-variable quantitative structure–activity relationship (QSAR) model was obtained using multiple linear regression methods. Four variables were found to account for both minimum effective dose (pMED) and the optimum effective dose (pOD) as shown in the following equations:

$$\text{pMED} = -0.208\ (\pm 0.040)\ ^1\chi^v + 2.11\ (\pm 0.289)\ ^4\chi^v_{pc}$$
$$- 0.338\ (\pm 0.030)\ S^T(-CH_3) \tag{8.7}$$
$$- 0.128\ (\pm 0.00949)\ S^T(\text{arom}) + 5.07\ (\pm 0.434)$$

$$r^2 = 0.90,\ s = 0.21,\ F = 70.0,\ n = 37$$

$$r^2_{press} = 0.85,\ s_{press} = 0.26$$

$$pOD = -0.207 \ (\pm 0.035) \ ^1\chi^v + 1.47 \ (\pm 0.254) \ ^4\chi^v_{pc}$$
$$- \ 0.264 \ (\pm 0.027) \ S^T(-CH_3) \tag{8.8}$$
$$- \ 0.0969 \ (\pm 0.00836) \ S^T(arom) + 5.40 \ (\pm 0.383)$$

$r^2 = 0.88, \ s = 0.19, \ F = 60.8, \ n = 37$

$r^2_{press} = 0.83, \ s_{press} = 0.23$

The $S^T(arom)$ variable is the combination of all the aromatic carbon E-State indices: $S^T(arom) = S^T(...CH...) + S^T(\cdot\cdot\overset{|}{C}\cdot\cdot) + S^T(\cdot\cdot\overset{:}{C}\cdot\cdot)$. The cross-validation statistics suggest that these models may be useful for prediction. The observed, calculated, and residual activity values are given in Table 8.8.

The combination of chi indices in this model describes aspects of the molecule as a whole which are indicated as important for activity prediction. The $^1\chi^v$ index encodes skeletal branching as well as molecular size. Since the carboquinones in this series do not vary greatly in size, skeletal variation is the most important aspect encoded. The negative coefficient indicates that greater branching, which produces a smaller index value, favors higher predicted activity. The $^4\chi^v_{pc}$ encodes more specific information about branching around a ring system (Kier, 1980). The greater the number of branch points on a ring, the greater the index value. Furthermore, the greater the adjacency of branch points, the smaller the index value. The positive value indicates that the greater the $^4\chi^v_{pc}$ value, the greater the predicted activity.

The two atom-type E-State indices encode information about two types of carbon atoms in the molecules. $S^T(-CH_3)$ is the sum of E-State values for $-CH_3$ groups. When the methyl group is near to atoms of low I values, its value is larger. Furthermore, the greater the number of methyl groups, the greater the value of $S^T(-CH_3)$. The positive value of the $S^T(-CH_3)$ coefficient indicates that the greater the value of $S^T(-CH_3)$, the greater the predicted activity. The $S^T(arom)$ index is the summation of all E-State values for aromatic carbons in the molecule. When electronegative atoms are attached to the aromatic carbon, the S value decreases. Hence, the negative coefficient indicates that smaller values of $S^T(arom)$ indicate higher activity.

In this study two forms of activity measurement were investigated—the minimum and the optimum dose. Both yielded good analysis, giving models with the same variables and similar coefficients. The quality of the statistics suggests that the models are useful for prediction in a drug design process.

TABLE 8.8 Antileukemic Activity as Medium Effective Dose (MED) and Optimum Dose (OD) for Carboquinones

	Substituent					Predicted
ID	R_1	R_2	pMED	Calculated	Residual	residual
1	C_6H_5	CH_5	4.33	4.17	0.16	0.34
2	CH_3	$(CH_2)_3C_6H_5$	4.47	4.73	−0.26	−0.33
3	C_5H_{11}	C_5H_{11}	4.63	4.54	0.09	0.14
4	$CH(CH_3)_2$	$CH(CH_3)_2$	4.77	5.20	−0.43	−0.69
5	CH_3	$CH_2C_6H_5$	4.85	4.96	−0.11	−0.13
6	C_3H_7	C_3H_7	4.92	5.06	−0.14	−0.16
7	CH_3	$CH_2OC_6H_5$	5.15	4.97	0.18	0.21
8	$CH_2OCON(CH_3)_2$	$CH_2OCON(CH_3)_2$	5.16	5.32	−0.16	−0.22
9	C_2H_5	C_2H_5	5.46	5.47	−0.01	−0.01
10	CH_3	$(CH_2)_2OCH_3$	5.57	5.70	−0.13	−0.13
11	CH_3O	CH_3O	5.59	5.76	−0.17	−0.22
12	CH_3	$CH(CH_3)_2$	5.60	5.57	0.03	0.04
13	C_3H_7	$CH(OCH_3)CH_2OCONH_2$	5.63	5.81	−0.18	−0.20
14	CH_3	CH_3	5.66	5.93	−0.27	−0.31
15	H	$CH(CH_3)_2$	5.68	5.56	0.12	0.13
16	CH_3	$CH(OCH_3)C_2H_5$	5.68	5.43	0.25	0.27
17	C_3H_7	$CH_2CH_2OCONH_2$	5.68	5.94	−0.26	−0.28
18	$CH_3OC_2H_4$	$CH_3OC_2H_4$	5.69	5.46	0.23	0.25
19	C_2H_5	$(CH_3O)CH_2OCONH_2$	5.76	5.71	0.05	0.05
20	CH_3	$(CH_2)_2OCONH_2$	5.78	5.90	−0.12	−0.12
21	CH_3	$(CH_2)_3dimer$	5.82	5.70	0.12	0.14
22	CH_3	C_2H_5	5.86	5.70	0.16	0.18
23	CH_3	$CH(OC_2H_5OCOCH_3)CH_2OCONH_2$	6.03	5.95	0.08	0.09
24	CH_3	$CH_2CH(CH_3)OCONH_2$	6.14	6.04	0.10	0.10
25	C_2H_5	$CH(CH_2OCONH_2)OCH_3$	6.16	6.02	0.14	0.15
26	CH_3	$CH(CH_2CH_3)CH_2OCONH_2$	6.18	6.24	−0.06	−0.07
27	CH_3	$CH(OC_2H_5)CH_2OCONH_2$	6.18	5.95	0.23	0.25
28	CH_3	$(CH_2)_3OCONH_2$	6.18	6.22	−0.04	−0.04
29	CH_3	$(CH_2)_2OCONH_2$	6.21	6.38	−0.17	−0.18
30	C_2H_5	$(CH_2)_2OCONH_2$	6.25	6.15	0.10	0.11
31	CH_3	C_2H_5OH	6.39	6.39	−0.00	−0.01
32	CH_3	$CH(CH_3)CH_2OCONH_2$	6.41	6.33	0.08	0.09
33	CH_3	$CH(OCH_3)CH_2OCONH_2$	6.41	6.26	0.15	0.16
34	H	$N(CH_2)_2$	6.45	6.66	−0.21	−0.25
35	$(CH_2)_2OH$	$(CH_2)_2OH$	6.54	6.84	−0.30	−0.36
36	CH_3	$N(CH_2)_2$	6.77	6.67	0.10	0.12
37	CH_3	$CH(OCH_3)CH_2OH$	6.90	6.27	0.63	0.67

TABLE 8.8 (*Continued*)

ID	Substituent		pOD	Calculated	Residual	Predicted residual
	R_1	R_2				
1	C_6H_5	C_6H_5	4.14	4.05	0.09	0.20
2	CH_3	$(CH_2)_3C_6H_5$	4.21	4.46	−0.25	−0.31
3	C_5H_{11}	C_5H_{11}	4.52	4.28	0.24	0.39
4	$(CH_3)_2CH_2$	$(CH_3)_2CH_2$	4.49	4.77	−0.28	−0.46
5	CH_3	$CH_2C_6H_5$	4.69	4.69	−0.00	−0.00
6	C_3H_7	C_3H_7	4.44	4.77	−0.33	−0.38
7	CH_3	$CH_2OC_6H_5$	4.71	4.70	0.01	0.01
8	$CH_2OCON(CH_3)_2$	$CH_2OCON(CH_3)_2$	4.85	4.77	0.08	0.12
9	C_2H_5	C_2H_5	5.09	5.13	−0.04	−0.04
10	CH_3	$(CH_2)_2OCH_3$	5.42	5.31	0.11	0.12
11	CH_3O	CH_3O	5.17	5.44	−0.27	−0.34
12	CH_3	$CH(CH_3)_2$	5.21	5.15	0.06	0.07
13	C_3H_7	$CH(OCH_3)CH_2OCONH_2$	5.07	5.28	−0.21	−0.23
14	CH_3	CH_3	5.36	5.53	−0.17	−0.20
15	H	$CH(CH_3)_2$	5.37	5.22	0.15	0.16
16	CH_3	$CH(OCH_3)C_2H_5$	5.33	5.04	0.29	0.32
17	C_3H_7	CH_2CH_2OCONH	5.23	5.44	−0.21	−0.23
18	$CH_3OC_2H_4$	$CH_3OC_2H_4$	5.31	5.09	0.22	0.25
19	C_2H_5	$CH(CH_3O)CH_2OCONH_2$	5.24	5.20	0.04	0.04
20	CH_3	$(CH_2)_2OCONH_2$	5.78	5.43	0.35	0.36
21	CH_3	$(CH_2)_3$dimer	5.39	5.33	0.06	0.07
22	CH_3	C_2H_5	5.37	5.33	0.04	0.05
23	CH_3	$CH(OC_2H_5OCOCH_3)CH_2OCONH_2$	5.39	5.39	0.00	0.00
24	CH_3	$CH_2CH(CH_3)OCONH_2$	5.79	5.51	0.28	0.29
25	C_2H_5	$CH(CH_2OCONH_2)OCH_3$	5.22	5.46	−0.24	−0.26
26	CH_3	$CH(CH_2CH_3)CH_2OCONH_2$	5.66	5.61	0.05	0.06
27	CH_3	$CH(OC_2H_5)CH_2OCONH_2$	5.22	5.41	−0.19	−0.20
28	CH_3	$(CH_2)_3OCONH_2$	5.93	5.67	0.26	0.27
29	CH_3	$(CH_2)_2OCONH_2$	5.75	5.82	−0.07	−0.07
30	C_2H_5	$(CH_2)_2OCONH_2$	5.48	5.62	−0.14	−0.15
31	CH_3	C_2H_5OH	5.79	5.87	−0.08	−0.09
32	CH_3	$CH(CH_3)CH_2OCONH_2$	5.71	5.71	−0.00	−0.00
33	CH_3	$CH(OCH_3)CH_2OCONH_2$	5.66	5.67	−0.01	−0.01
34	H	$N(CH_2)_2$	6.19	6.11	0.08	0.10
35	$(CH_2)_2OH$	$(CH_2)_2OH$	6.05	6.19	−0.14	−0.17
36	CH_3	$N(CH_2)_2$	6.21	6.04	0.17	0.20
37	CH_3	$CH(OCH_3)CH_2OH$	5.75	5.72	0.03	0.04

VIII. TOXICITY OF AMIDE HERBICIDES

Of great interest and value are the abilities to model and to predict the toxicities of molecules of commerce. Nowhere is this recognized more than in the agrochemical industry. The huge volume of herbicides and pesticides in use demands that a best effort be made to understand and to predict, as far as possible, the toxic consequences of their use. Perhaps more attention is currently being paid to the modeling and the prediction of toxicity among existing agents than to the development of new agents. It is in this realm of science and technology that structure descriptors come into their own.

A recent example of such modeling and prediction is provided in a study by Gough and Hall (1998b). Acute, oral toxicities of 50 amides (Scheme VII) in rats, reported (Jager, 1983; Thomson, 1983; Bohmont et al., 1987; Zakarya et al., 1996) and converted to $-\log LD_{50} = pLD_{50}$, were explored using E-State and topological indices.

$$\text{(VII)}$$

The following were determined to be the best five variables:

$$pLD_{50} = 0.189\ (\pm 0.039)\ ^2\chi^v$$
$$+\ 0.364\ (\pm 0.070)\ S(N\text{-}1)$$
$$-\ 0.071\ (\pm 0.020)\ SH^T(other)$$
$$-\ 0.700\ (\pm 0.094)\ S^T(=C<)$$
$$-\ 0.979\ (\pm 0.429)\ SH^T(-OH)$$
$$-\ 0.304\ (\pm 0.204)$$

$$r^2 = 0.75,\ s = 0.23,\ F = 26,\ n = 50$$

$$r^2_{press} = 0.65,\ s_{press} = 0.27$$

The intercorrelation of the parameters is very low, with r^2 ranging from 0.003 to 0.35. The observed and calculated values are shown in Table 8.9. The regression model was used to predict the toxicities of nine amides not included in the analysis set. These predictions are shown in Table 8.10. For these predicted toxicities the mean absolute error (MAE) is 0.27 for the 12 data values available.

The QSAR model discussed previously may be interpreted in terms of the molecular structure of the amides because of the structure-based character of the indices in the model. The percent of calculated toxicity has been computed for each of the variables as follows: $^2\chi^v$, 37.2%; $S(\text{N-1})$, 34.4%; $SH^T(\text{other})$, 15.5%; $S^T(=C<)$, 12.5%; $SH^T(-\text{OH})$, 0.4%.

The $^2\chi^v$ variable is related to the degree of skeletal branching, increasing as the amount of branching increases within a molecule. This index also increases with the number and complexity of rings. Furthermore, the index is greater for family members higher in the periodic table such as chlorine and bromine compared to fluorine. The positive coefficient indicates that a larger index value is related to greater toxicity.

The atom E-State index for the amide nitrogen atom $S(\text{N-1})$ contributes about one-third of the calculated toxicity. In this data series this atom is either a $-\text{NH}-$ or a $>\text{N}-$ group. Atoms near the amide nitrogen have the greatest effect on its E-State value. Atoms with low I values tend to enrich $S(\text{N-1})$. For the 10 most toxic amides, $S(\text{N-1}) > 2.0$. Such an observation can be useful in analyzing toxicity or in designing new amide herbicides. The positive coefficient of $S(\text{N-1})$ indicates that the larger its value, the greater the calculated toxicity.

The third index in Eq. (8.9), $SH^T(\text{other})$, represents the portion of a molecule which is nonpolar. The actual index is calculated only for the hydrogen atoms, based on the relative valence state electronegativity of all other atoms or groups in the molecule. These hydrogen atoms reside in groups which are less acidic than an acetylenic $\equiv\text{CH}$. The more nonpolar these hydrogen atoms and the greater their number, the larger the $SH^T(\text{other})$ value. The negative coefficient on $SH^T(\text{other})$ indicates that the smaller the index, the greater the calculated toxicity. Of similar importance to the $SH^T(\text{other})$ index is the atom-type E-State index for trigonal carbon atoms with one double bond and two single bonds, $S^T(=C<)$. Because this atom type has the smallest intrinsic state I value, all such atom E-State values are less than the intrinsic state (1.667). The negative value for $S^T(=C<)$ indicates that the larger the index value, the smaller the calculated toxicity.

The last index in Eq. (8.9), $SH^T(-\text{OH})$, contributes to the model in a different manner. $SH^T(-\text{OH})$ stands for the sum of hydrogen E-State values for all $-\text{OH}$ groups in the molecule. It represents a very polar hydrogen atom. The negative coefficient on $SH^T(-\text{OH})$ indicates that the greater the index, the lesser the predicted toxicity. In this data set, however, there are only two compounds which contain the OH group. The $SH^T(-\text{OH})$ index was added so that the model adequately accounts for these two compounds. If the index is omitted, the regression statistics are the same. However, the predicted residuals for the OH-containing molecules are high; thus, inclusion of $SH^T(-\text{OH})$ accounts for the OH contribution to toxicity for these two compounds. Because only 2 of the 50 compounds contain OH, it is anticipated that

TABLE 8.9 Toxicity against Male Rats [-log(LD$_{50}$)] for a Set of Amide Herbicides

Observation	Substituent R$_1$	R$_2$	R$_3$	pLD$_{50}$[a]	Calculated[b]	Residual[c]	Predicted residual[d]
1	3-NHCOOC$_2$H$_5$—C$_6$H$_4$—O—	H	—C$_6$H$_5$	2.42	1.74	0.68	0.78
2	2,6,-diC(CH$_3$)$_3$, 4-CH$_3$—C$_6$H$_2$—O—	H	—CH$_3$	2.10	2.08	0.02	0.03
3	Cl$_2$CH—	H	(chlorinated naphthoquinone structure)	1.67	1.47	0.21	0.22
4	C$_2$H$_5$NHCOCH(CH$_3$)O—	H	—C$_6$H$_5$	1.67	1.67	0.00	-0.00
5	C$_6$H$_5$NHCOOCH(CH$_3$)—	H	—C$_2$H$_5$	1.67	1.63	0.04	0.04
6	3,5,-diCl—C$_6$H$_5$—	H	—HC=CC(CH$_3$)$_2$	1.59	1.50	0.09	0.10
7	n-C$_3$H$_7$CH(CH$_3$)—	H	3,4-diCl—C$_6$H$_5$—	1.58	1.18	0.41	0.46
8	3,5-diCl—C$_6$H$_3$—	H	—CHC=C(CH$_3$)$_2$	1.51	1.43	0.08	0.08
9	C$_2$H$_5$—	H	HOOCCH$_2$O—	1.46	1.26	0.20	0.34
10	2-COOH—C$_6$H$_4$—	H	Napthyl-	1.45	1.59	-0.14	-0.49
11	(CH$_3$)$_2$CHO—	H	C$_6$H$_5$—	1.45	1.33	0.12	0.13
12	4-NHCOOCH$_3$—C$_6$H$_4$—O—	H	3-CH$_3$—C$_6$H$_4$—	1.43	1.71	0.28	0.33
13	CH$_3$O—	(CH$_3$)$_2$CHCH$_2$—	4-NH$_2$—C$_6$H$_4$—SO$_2$—	1.34	1.43	-0.10	-0.11
14	C$_2$H$_5$S—	C$_2$H$_5$—	(CH$_3$)$_2$CH=CH—	1.33	1.17	0.16	0.19
15	C$_2$H$_5$S—	C$_2$H$_5$—	Cyclohexyl-	1.22	1.05	0.17	0.19
16	(CH$_3$)$_3$CCH(Br)—	H	C$_6$H$_5$—C(CH$_3$)$_2$—	1.20	1.44	-0.24	-0.34
17	C$_6$H$_5$—	(CH$_3$)$_2$N—CO—	3,4-diCl—C$_6$H$_3$—	1.17	0.98	0.19	0.23
18	C$_6$H$_5$—	CH$_3$OCOCH(CH$_3$)—	3-Cl,4-F—C$_6$H$_3$—	1.17	0.88	0.29	0.32
19	(CH$_2$)(CH$_3$)CH—	H	3,4-diCl—C$_6$H$_3$—	1.14	0.90	0.24	0.27
20	(imidazolidinone structure)	H	i-C$_3$H$_7$CH$_2$—	1.13	1.35	-0.22	-0.23
21	4-NHCOO—i—C$_3$H$_7$—C$_6$H$_4$—O—	C$_2$H$_5$—	C$_6$H$_5$—	1.07	1.30	-0.23	-0.27
22	HC≡CCH(CH$_3$)O—	H	3-Cl—C$_6$H$_4$—	1.03	1.37	-0.34	-0.37
23	ClCH$_2$—	n-C$_4$H$_9$OCH$_2$—	2,6-diC$_2$H$_5$—C$_6$H$_3$—	1.02	0.92	0.11	0.11
24	3-CH$_3$—C$_6$H$_4$—	C$_2$H$_5$—	C$_2$H$_5$—	1.02	0.69	0.33	0.34

186

No.				pLD₅₀[a]	Calc.[b]	Resid.[c]	Pred. resid.[d]

Structure (compounds 25–31): H₃C— on a thiazoline ring (N, S) bearing Cl.

No.				pLD_{50}[a]	calc.[b]	resid.[c]	pred.[d]
25	C₂H₅—	H	Cl	1.00	0.83	0.17	0.20
26	ClCH₂—	CH₃OCH₂CH(CH₃)—	2-CH₃,6-C₂H₅—C₆H₃—	0.99	0.87	0.12	0.13
27	CH₂Cl—	CH₂CHCH₃—	CH₂CHCH₃—	0.98	0.83	0.15	0.17
28	n-C₃H₇S—	n-C₃H₇—	n-C₃H₇—	0.94	0.89	0.06	0.06
29	C₂H₅S—	n-C₃H₇—	n-C₃H₇—	0.94	0.86	0.08	0.08
30	3,4-diCl—C₆H₃—CH₂O—	H	CH₃—	0.92	1.25	−0.33	−0.34
31	(CH₃)₂C=NO—	H	C₆H₅—	0.90	0.95	−0.04	−0.05

Structure (compounds 32–50): four-membered ring, N—N=CH₂.

No.				pLD_{50}[a]	calc.[b]	resid.[c]	pred.[d]
32	ClCH₂—	2,6-di(C₂H₅)C₆H₃—	2,6-diCH₃—C₆H₃—	0.89	0.80	0.09	0.09
33	CH₂Cl—	CH₃OCH₂—	CH₃CH₂OCOCH₂—	0.87	1.13	−0.26	−0.28
34	CH₂Cl	H	2,6-C₆H₃—	0.82	0.80	0.02	0.02
35	C₂H₅—	2,6-diCH₃,C₆H₃—	3,4diCl(C₆H₃)—	0.81	1.06	−0.25	−0.28
36	CH₂CH₂Cl—	C₂H₅—	CH₃OCH₂CH₂—	0.80	0.76	0.04	0.04
37	C₃H₇S—	CH₃OCH₂—	C₄H₉—	0.77	0.87	−0.11	−0.11
38	ClCH₂—	HC≡CCH(CH₃)—	2,6-diC₂H₅—C₆H₃—	0.75	0.80	−0.06	−0.06
39	ClCH₂—	C₂H₅—	C₆H₅—	0.73	0.69	0.04	0.04
40	4-ClC₆H₄—CH₂S—	H	C₂H₅—	0.70	0.82	−0.12	−0.13
41	CH₂ClCH₂C≡CCH₂OCH₂—	C₂H₅—	3-Cl—C₆H₄—	0.70	1.11	−0.41	−0.46
42	2-Cl—C₆H₄—CH₂—S—	C₂H₅—	C₂H₅—	0.63	0.80	−0.17	−0.18
43	C₆H₅—	C₂H₅OCOCOCH(CH₃)—	3,4,diCl—C₆H₃—	0.63	0.84	−0.21	−0.31
44	(C₆H₅)₂CH—	CH₃—	CH₃—	0.62	0.64	−0.02	−0.02
45	ClCH₂—	CH₂=CHCH₂—	CH₂=CHCH₂—	0.62	0.56	0.06	0.06
46	CH₂C(Cl)CH₂S—	C₂H₅—	C₂H₅—	0.61	0.45	0.16	0.18
47	(C₆H₅)₂CH—	CH₃—	CH₃—	0.61	0.64	−0.04	−0.04
48	Cl₂H—	CH₂=CHCH₂—	CH₂=CHCH₂—	0.56	0.83	−0.27	−0.30
49	SC₂H₅—	C₂H₅—	C₂H₅—	0.39	0.72	−0.32	−0.35
50	CHClCClCH₂S—	(CH₃)₂CH—	(CH₃)₂CH—	0.16	0.30	−0.14	−0.17

---

[a] Toxicity taken as pLD₅₀ = −log(LD₅₀) from Jager (1983), Thomson (1983), Bohmont et al. (1987), Zakaya et al. (1996), Yalkowsky and Banerjee (1992), Meylan et al. (1996), and Klopman et al. (1992) and converted to molar basis.

[b] Calculated toxicity from Eq. (8.4). (see text)

[c] Residual = pLD₅₀ − calculated.

[d] Predicted residual computed by the leave-one-out method.

TABLE 8.10 Prediction Test Set for the Toxicity Model

Observation	Substituent			Toxicity (mg/kg)	pLD$_{50}$[a]	Calculated[b]	Residual[c]
	R$_1$	R$_2$	R$_3$				
51	n-C$_3$H$_7$CH(CH$_3$)—	H	3-Cl,4-CH$_3$—C$_6$H$_3$—	>10,000	0.53	1.21	-0.68
52	2,6-diOCH$_3$—C$_6$H$_3$—	H	H$_3$C—CH$_2$ / H$_3$C—C—CH$_2$— (5-methylisoxazol-3-yl)	>10,000	1.48	1.54	-0.06
53	CH$_3$O—	H	3,4-diCl—C$_6$H$_3$—[d]	525	0.38	1.30	-0.92
				4,197	1.28	1.30	-0.02
54	C$_6$H$_5$—	i-C$_3$H$_7$OCOCH(CH$_3$)—	3-Cl,4-F—C$_6$H$_3$—	>4,000	1.05	0.87	0.18
55	n-C$_3$H$_7$C(CH$_3$)$_2$—	H	4-Cl—C$_6$H$_4$—	>4,000	1.22	1.34	-0.12
56	CH$_3$—	H	2,4-diClCH$_3$,6-NHSO$_2$CF$_3$—C$_6$H$_2$—	>4,000	1.16	1.58	-0.42
57	i-C$_3$H$_7$O—	H	3-Cl—C$_6$H$_4$—	5,000–7,500	1.25–1.37	1.49	-0.18[f]
58	Cl$_2$C=C(Cl)CH$_2$S—	i-C$_3$H$_7$—	i-C$_3$H$_7$—	1,675–2,165	0.74–0.85	1.12	-0.33[f]
				710	0.53	0.76	-0.23
59	ClCH$_2$—	iCH$_7$	C$_6$H$_5$—[e]	1,200	0.75	0.76	-0.01
				1,200–1,475	0.75–0.84	0.76	0.04[f]

[a] Toxicity taken as pLD$_{50}$ = -log(LD$_{50}$) and converted to molar basis.

[b] Calculated toxicity from Eq. (8.4) (see text).

[c] Residual = pLD$_{50}$ − Calculated.

[d] Two data entries are available in the references.

[e] Three data entries are available in the references.

[f] Residual is the difference between the average of observed range values and calculated values.

prediction of OH-containing molecules is less reliable than that of the remaining compounds.

The equation model provides ample information for the development of conclusions about structural attributes influencing the toxicity. This QSAR model adequately accounts for the amide herbicide toxicity within the limitations of the uncertainties in the measured toxicity data.

IX. AQUEOUS SOLUBILITY OF DRUG MOLECULES USING ARTIFICIAL NEURAL NETWORKING

An important aspect of the bioavailability of a drug is its aqueous solubility. Several investigators have attempted theoretical prediction of aqueous solubility (Yalkowsky and Banerjee, 1992; Meylan et al., 1996; Klopman et al., 1992), but Lipinski et al. (1997) discussed these approaches with some reservations as to their general applicability. In Chapter 5, Section X, we provided an example in which solubility was successfully predicted for one set of drug molecules using the E-State indices in a neural network approach. Here, we present a study of a larger, more diverse data set using the same approach.

Huuskonen et al. (1998) assembled a set of 211 drugs and related compounds from several literature sources. In a previous work Huuskonen et al. (1997) presented three studies of solubility prediction using neural networks based on the E-State indices using the neural network approach. The larger study is an extension of the earlier work in which several molecular connectivity chi indices are combined with the E-State indices for the representation of molecular structure.

The solubility is given as log(S) in which S is expressed as moles/liter. The log S values range from 0.55 (ethambutol) to -5.60 (thioridazine). The data set was divided into a training set of 160 compounds and 51 test compounds on a random basis. Atom-type E-State indices and molecular connectivity chi indices were computed along with kappa shape indices, the count of hydrogen bond acceptors and donors, and an indicator variable for aromatic compounds. Preliminary analysis permitted the selection of a set of 32 variables as shown in Table 8.11.

The preliminary neural network analysis indicated that a 32:5:1 architecture was optimal. Sensitivity analysis was conducted to determine the most important variables in the network. The least important variables were eliminated. Sensitivity was estimated by summing all the network weights related to a given input. Furthermore, a noise input variable was added to give the network a choice between the noise and a significant variable. Based on this anal-

TABLE 8.11   Structure Variables Used in the Neural
Network Model of Aqueous Solubility

Atom-type indices[a]			
Symbol	Atom type	Symbol	Atom type
$S^T(-CH_3)$	$-CH_3$	$S^T(>N-)$	$>N-$
$S^T(-CH_2-)^b$	$-CH_2-$	$S^T(=N-)^b$	$=N-$
$S^T(>CH-)$	$>CH-$	$S^T(...N...)$	$...N...$
$S^T(>C<)$	$>C<$	$S^T(-OH)$	$-OH$
$S^T(=CH_2)$	$=CH_2$	$S^T(-O-)$	$-O-$
$S^T(>CH-)$	$=CH-$	$S^T(=O)$	$=O$
$S^T(>C=)$	$=C<$	$S^T(-F)^b$	$-F$
$S^T(...CH...)$	$...CH...$	$S^T(-SH)^b$	$-SH$
$S^T(...C...)$	$...C...$	$S^T(-S-)$	$-S-$
$S^T(-NH_2)^b$	$-NH_2$	$S^T(=S)^b$	$=S$
$S^T(-NH-)^b$	$-NH-$	$S^T(-Cl)^b$	$-Cl$

Topological indices[a]	
Symbol	Index type
$^2\chi$	Path 2 simple chi index
$^3\chi_c$	Cluster 3 simple chi index
$^6\chi^v_{ch}$	Chain 6 valence chi index
$^7\chi^v_{ch}{}^b$	Chain 7 valence chi index
$^9\chi^v_{ch}$	Chain 9 valence chi index
$^{10}\chi^v_{ch}$	Chain 10 valence chi index
$^3\kappa_\alpha$	Kappa alpha 3 shape index
AR	Aromaticity indicator[c]
numhbd	Number of hydrogen bond donors
numhda	Number of hydrogen bond acceptors

[a] For atom-type E-State definition see Chapter 4.

[b] These indices were eliminated on the basis of sensitivity analysis.

[c] For aromatic compounds, AR = 1; AR = 0 otherwise.

ysis, nine variables were removed, resulting in a 23:5:1 network for the final analysis. The deleted variables are indicated in Table 8.11.

The statistical results of the network modeling are as follows:
For the training set: $r^2 = 0.90$, $s = 0.46$, $n = 160$; for the test set: $r^2 = 0.86$, $s = 0.53$, $n = 51$. There were only 4 compounds in the training set and 4 in the test set with residuals ≥ 2 standard deviations. Klopman *et al.* (1992) considered a set of 21 eclectic compounds as a test of their group contribution method. The current neural network model gave the same standard deviation as that of Klopman *et al.*: $s = 1.25$. In general, this current method was of the

same quality; however, there were two pesticides that had large deviations. For the complete list of the observed, calculated, and residual solubilities, the reader is referred to Huuskonen et al. (1998).

This study indicates the potential for successful prediction of aqueous solubility for organic compounds using a combination of indices with the neural network formalism. However, the Lipinski comment (1997) on the importance of the nature of the training set should be taken seriously. Currently, it is not clear how big a set is needed and what criteria should be used in selecting the compounds. In this case, in which the training set is largely drug-like, pesticides are not well predicted in the test set. Based on molecular structure considerations, such a result is not surprising. It appears, however, that when a large data set is compiled, the E-State indices in combination with other topological indices may well prove useful in the prediction of solubility.

X. INHIBITION OF MONOAMINE OXIDASE BY PHENETHYLAMINES: RELATION TO FREE VALENCE

In Chapter 3, Section IV A, we discussed the concept of free valence expressed as an insight into the nature of the E-State indices. That section dealt primarily with the free valence for carbon atoms because the original work on free valence did not deal with heteroatoms. We examine a data set of phenethylamines (Scheme VIII) which are inhibitors of monoamine oxidase (MAO) as reported by Dunmire and Hall (1998).

(VIII)

The partial least squares (PLS) method is used to develop a model of the structure–activity relation based on a combination of E-State indices along with molecular connectivity chi and kappa shape indices. A model based on an extended free valence approach is revealed.

Biological activity, as in vitro inhibition against MAO, was taken from Hasagawa et al. (1996). The list of compounds and their observed activities, pIC50 = $-\log(IC_{50})$, are given in Table 8.12. The activity varies from 4.90 for the 4-isopropyl compound to 8.79 for the 2,6-dichloro-4-dimethylamino compound. To provide for substitution symmetry around the phenyl ring, the

TABLE 8.12   Inhibition of MAO by Phenethylamines

Observed	Name		Substituent		pIC$_{50}$	Calculated	Residual
1	FLA-299		4-CH(CH$_3$)$_2$		4.90	5.19	−0.29
2	FLA-289		4-N(CH$_3$)$_2$		5.43	5.41	0.02
3	FLA-558	2-F	4-N(CH$_3$)$_2$		5.92	6.16	−0.24
4	FLA-314	2-Cl	4-N(CH$_3$)$_2$		6.68	6.72	−0.04
5	FLA-405	2-Br	4-N(CH$_3$)$_2$		6.66	6.22	0.44
6	FLA-336	2-CH$_3$	4-N(CH$_3$)$_2$		5.57	5.65	−0.08
7	Amiflamine(+)	2-CH$_3$	4-N(CH$_3$)$_2$		6.10	5.86	0.24
8	FLA-336(−)	2-CH$_3$	4-N(CH$_3$)$_2$		5.52	5.44	0.08
9	FLA-365	2-Cl	4-N(CH$_3$)$_2$	6-Cl	7.89	7.95	−0.06
10	FLA-463	2-Cl	4-N(CH$_3$)$_2$	α-CH$_3$	4.92	4.86	0.06
11	FLA-717	2-CH$_3$	4-N(CH$_3$)$_2$	α-CH$_3$	4.92	4.86	0.06
12	FLA-384	3-CH$_3$	4-N(CH$_3$)$_2$		5.10	5.27	−0.17
13	NBF-003(+)	2-CH$_3$	4-N(CH$_3$)$_2$	5-Br	5.96	6.18	−0.22
14	FLA-727		4-N(CH$_3$)$_2$		6.26	6.34	−0.08
15	FLA-788(+)	2-CH$_3$	4-NHCH$_3$		6.89	6.81	0.08
16	NBF-008(+)	2-CH$_3$	4-NHCH$_3$	5-Br	7.05	7.14	−0.09
17	FLA-334		4-NH$_2$		5.14	5.23	−0.09
18	FLA-668	2-CH$_3$	4-NH$_2$		5.80	5.47	0.33
19	FLA-668(−)	2-CH$_3$	4-NH$_2$		5.12	5.26	−0.14
20	NBF-021(+)	2-CH$_3$	4-NH$_2$	5-Br	6.10	6.00	0.10

E-State indices for *ortho* position 2 and 6 were combined: $S(C\text{-}2,6) = S(C\text{-}2) + S(C\text{-}6)$. A similar approach was followed for the *meta* position and for the hydrogen E-State indices for both sets of positions. In addition, an indicator variable was included to encode stereochemistry: Stereo $= +1$ for $+$, -1 for $-$, and 0 for racemic mixture. The variables are listed in Table 8.13.

The data set was subjected to PLS analysis using the SAS Statistical System (The SAS Institute, P.O. Box 8000, Cary, NC 27511-8000). A model based on three latent variables yielded the following statistics:

$$r^2 = 0.94, \ s = 0.21, \ n = 20; \ \text{c.v.} \ r^2 = 0.92, \ s = 0.25, \ n = 20$$

The observed, calculated, and residual activities are given in Table 8.12. Only one residual is slightly ≥ 2 standard deviations. The cross-validated standard deviation is also only slightly larger than the direct regression error, indicating a reliable model for prediction. A plot of calculated versus observed activity shows no systematic trends. The PLS loadings on the variables are given in Table 8.13; these are rank ordered on the magnitude of the loadings.

Free valences were calculated for each atom in the molecule using the AM1 method in AMPAC (SemiChem, PO Box 1649, Shawnee Mission, KS 66222). Based on the bond orders calculated in the AM1 molecular orbital method, free

TABLE 8.13 Variable Loadings: PLS Modeling of
Phenethylamine Inhibition of MAO

Descriptor No.	Variable	Weight	%
1	$HS(C\text{-}7)$	2.92	6.53
2	$HS(C\text{-}10)$	2.88	6.43
3	$HS(C\text{-}3,5)$	2.81	6.28
4	$HS(N\text{-}9)$	2.76	6.17
5	$S(C\text{-}2,6)$	2.40	5.35
6	$S^T(...CH...)$	2.28	5.09
7	$HS(C\text{-}2,6)$	2.25	5.04
8	$S^T(-CH_2-)$	2.25	4.91
9	$S(C\text{-}7)$	2.20	4.91
10	$S(C\text{-}1)$	1.88	4.21
11	$HS(C\text{-}8)$	1.81	4.04
12	$S(C\text{-}8)$	1.59	3.55
13	$S^T(-Cl)$	1.47	3.28
14	$^4\chi^v_{pc}$	1.44	3.21
15	$S(C\text{-}10)$	1.43	3.19
16	$^2\kappa_\alpha$	1.33	2.96
17	$S^T(>CH-)$	1.29	2.89
18	$^1\chi^v$	1.23	2.76
19	$S^T(-Br)$	1.20	2.69
20	$S(C\text{-}3,5)$	1.17	2.62
21	$S^T(-NH-)$	1.04	2.32
22	$S^T(>N-)$	0.90	2.00
23	Stereo	0.88	1.97
24	$^3\kappa_\alpha$	0.79	1.75
25	$S^T(-F)$	0.78	1.74
26	$S^T\left(\begin{array}{c}\mid\\ \cdot\cdot C \cdot\cdot\end{array}\right)$	0.74	1.65
27	$S^T(-NH_2)$	0.31	0.69
28	$S^T(-CH_3)$	0.27	0.61
29	$S(N\text{-}9)$	0.27	0.61
30	$S(C\text{-}4)$	0.21	0.46

valences were obtained for each atom as $F_k = N_{max} - N_k$, where N_k is the sum of bond orders at an atom. For each lone pair of electrons, 1.0 was added to N_k. A data set was developed in parallel to the one based on the E-State. PLS analysis was also run on this free valence data set, and the PLS results indicated that no useful model could be found. Although this extension of the free valence concept to heteroatoms could probably be improved, these results indicate that the mere use of bond orders to represent atoms in molecules is significantly inferior to the use of the E-State indices.

Examination of the loadings given in Table 8.13 indicates the great importance of the hydrogen atom E-State indices—5 of the first 7 indices in Table 8.13. Three of the most important variables refer to the side chains: $HS(C-7)$, $HS(N-9)$, and $HS(C-10)$. Of the first 15 most important indices, approximately one-third refer to side chain carbon and hydrogen atoms and approximately one-fourth to phenyl ring carbon atoms. These results indicate the relative importance of the side chain. In this model the electronic and steric effects of the various heteroatoms are expressed indirectly in the E-State values of both the E-State indices and the hydrogen E-State indices. For example, substitution at the 2,6 *ortho* positions produces effects in the side chain. These effects are encoded in the E-State and hydrogen E-State indices of the side chain atoms.

The Stereo variable has a low weight in this model; however, the effects of stereochemistry (at position 8) are encoded in the E-State indices $S(C-8)$ and $HS(C-8)$, indicating a role of importance for stereochemistry. The whole molecule chi and kappa indices have low weights, indicating that whole molecule effects are less important here than are more localized or regional effects. When the chi and kappa indices are eliminated from the model, the standard error is only slightly reduced.

Overall, the PLS model based on the atom-level and atom-type E-State indices appears reliable. The method is applied in a straightforward manner. The indices are readily calculated from connection tables, and the statistical analysis is accomplished with standard software packages.

XI. DIRECT QSAR MODELING OF PERCENT EFFECT DATA

In the usual QSAR modeling investigation, the biological data consist of a computed measure of the biological response. This computed quantity is obtained from an analysis of the log(dose)–response curve and reported as an IC_{50}, EC_{50}, LD_{50}, etc. There are several situations, however, in which such response values are not available, especially in the earliest stages of a design project. Insufficient data are available to establish the full dose–response curve. There are many reasons for this situation, including economical measures to limit costs, limited solubility, insufficient information to set the concentration values for the screening protocols, and measures to save experimental animals (Wiese and Schaper, 1993).

In an effort to deal with this situation, two major approaches have been attempted. In the first, the observed effects are classified into arbitrarily defined activity classes and these classes are used in pattern recognition methods or adaptive least squares (Schaper and Saxena, 1991). In the other approach a logit transformation is used: logit% = log[%/(100 − %)]; these values are analyzed with standard multiple linear regression, often using physicochemical parameters as independent variables (Schaper and Saxena, 1991; Schaper, 1987). Wiese (1991) applied neural network analysis directly to the percent effect data.

A. CHLOROMETHANESULFONAMIDE ACARICIDAL ACTIVITY AGAINST CITRUS RED MITES

In this study, we apply neural network methodology to dose-dependent data using the E-State indices as the structure representation. The data set was taken from Tamaru *et al.* (1988) as 43 observations on 27 compounds. The compounds are chloromethanesulfonamides (Scheme IX) and the activity is acaricidal activity against citrus red mites (*Panonuchus citrii* McGregor) as shown in Table 8.14.

(IX)

Preliminary analysis using multiple linear least squares was performed with percent effect as the dependent variable and the log dose as one of the independent variables. The atom-level E-State indices for the eight atoms in the common molecular scaffold were entered along with appropriate atom-type E-State indices and a selection of chi and kappa shape indices. For any intercorrelated pairs of variables ($r > 0.80$), one was eliminated. Fourteen variables appeared in several of the linear equations, although none of the models had satisfactory statistics. These 14 variables were then submitted to neural network analysis: $S(C\text{-}2)$, $S(S\text{-}3)$, $S(O\text{-}4)$, $S(N\text{-}6)$, $HS(C\text{-}2)$, $HS(N\text{-}6)$, $S^T(\text{—O—})$, $S^T(\text{—Cl})$, $S^T(\text{—NH—})$, $SH^T(\text{—NH—})$, $^2\chi^v$, $^4\chi_P^v$, $^7\chi_P^v$, and $^2\kappa_\alpha$ in addition to the concentration at which the percent effect is obtained.

B. NEURAL NETWORK ANALYSIS

The previously discussed data set was used with an experimental approach to neural network analysis under development at Parham Analysis (M. Parham, Parham Analysis, 6 Reuben Duren Way, Bedford, MA 01730-1666). All inputs

TABLE 8.14   Chloromethanesulfonamides and Percent Effect

ID[a]	Substituent R_1	R_2	log Concentration[b] -log C	% Effect Observed	Calculated	Residual
1	H	H	2.41	99	99.0	0.0
2			2.11	99	99.0	0.0
3			1.81	100	100.0	0.0
4	Me—	H	2.46	91	91.0	0.0
5			2.16	91	91.0	0.0
6			1.86	100	100.0	0.0
7	Et—	H	2.20	88	88.0	0.0
8	Pr—	H	2.23	82	82.0	0.0
9[a]	i—Pr—	H	2.27	82	82.0	4.0
10	Bu—	H	2.27	82	82.0	0.0
11	t-Bu—	H	2.38	72	72.0	0.0
12	Bu—CH(Et)CH$_2$—	H	2.38	75	75.0	0.0
13	OH—CH$_2$CH$_2$—	H	2.54	0	0.0	0.0
14[a]			2.24	61	58.6	2.4
15[a]			1.94	76	76.6	−0.6
16[a]	EtO—CH$_2$—CH$_2$—	H	2.30	68	83.0	−15.0
17	MeO—(CH$_2$)$_3$—	H	2.61	75	75.0	0.0
18			2.30	83	83.0	0.0
19			2.00	99	99.0	0.0
20	EtSCH$_2$CH$_2$—	H	2.34	64	64.0	0.0
21	PrSCH$_2$CH$_2$—	H	2.37	0	0.0	0.0
22	PhSCH$_2$CH$_2$—	H	2.42	0	0.0	0.0
23	PhCH$_2$SCH$_2$CH$_2$—	H	2.75	0	0.0	0.0
24			2.45	0	0.0	0.0
25			2.15	0	0.0	0.0
26[a]	PhCH$_2$—	H	2.64	0	0.0	0.0
27[a]			2.34	75	75.6	−0.6
28			2.04	89	89.0	0.0
29	CH$_2$=CH—CH$_2$—	H	2.23	62	62.0	0.0
30	Ph	H	2.31	0	0.0	0.0
31	4-Cl—Ph	H	2.38	4	4.0	0.0
32	4-Me—Ph	H	2.34	0	0.0	0.0
33	4-MeO—Ph	H	2.37	0	0.0	0.0
34[a]	Me	Me	2.50	82	87.0	−5.0
35			2.20	87	87.0	0.0
36			1.90	100	100.0	0.0
37	Et	Et	2.57	0	0.0	0.0
38			2.27	0	0.0	0.0
39			1.97	0	0.0	0.0
40	i-Pr	i-Pr	2.33	0	0.0	0.0
41	i-Bu	i-Bu	2.38	0	0.0	0.0
42[a]	Ph	Me	2.34	33	0.6	32.4
43	NC—CH$_2$—CH$_2$—	Et	2.32	0	0.0	0.0

[a] Included in the testing set.

[b] Molar concentration at which the percent effect was measured.

to the network are ranked in importance based on an initial assessment of the relative importance of each variable in the model. The network is allowed to operate with a large number of hidden neurons up to the number of rows of data (number of observations). The algorithm may prune hidden nodes from the network, starting from the maximum number until an optimum network solution is found based on the r^2 values of actual experimental values versus predicted values. In this case, the data organization inclusion criteria mandates that certain data be included in the training set: (i) all maximum or minimum values and (ii) the most significant input data values.

For the chloromethanesulfonamide data, no individual network was clearly superior. Therefore, a consensus of networks was used to identify areas of agreement between the network predictions (across five networks) based on cross-validation of those values which are neither maximum nor minimum. This procedure improved the accuracy of the overall network output above that of an individual network prediction. In each network, agreement between the validated prediction for all training set data is considered in selection of each prediction. With a small compact database using the E-state indices, excellent agreement between actual percent effect and the predicted percent effect values was obtained.

The network analysis described previously identified 6 of the 14 variables with high statistical weight; that is, the important variables for the relation between percent effect and structure. These 6 structure variables, in addition to the concentration, in order of relative importance are as follows: $S(O\text{-}4)$, $^7\chi_P^v$, $HS(N\text{-}6)$, concentration, $S^T(\!-\!O\!-\!)$, and $HS(C\text{-}2)$. For this data set the statistical analysis indicates excellent results:

Total data set: $r^2 = 0.983$
Training data set: $r^2 = 0.999$
Test set: $r^2 = 0.881$

The MAE = 1.4. The largest residual is obtained for observation No. 42 (32.4), which suggests that the network cannot learn this point. The plot of calculated versus observed percent effect is shown in Fig. 8.1. The calculated values are given in Table 8.14 along with the residual values. It is clear that these results are excellent; the agreement is very close.

These results appear to be significantly better than those obtained by Wiese and Schaper (1993). These authors found four observations with large deviations (Nos. 14, 20, 25, and 42). Their standard deviation is reported as 15.2, whereas the MAE is 9.3—much greater than the 1.4 found in our study. The difference in the results is a combination of the use of the Parham method, the effectiveness of the E-State structure representation, and the number of variables included in the analysis.

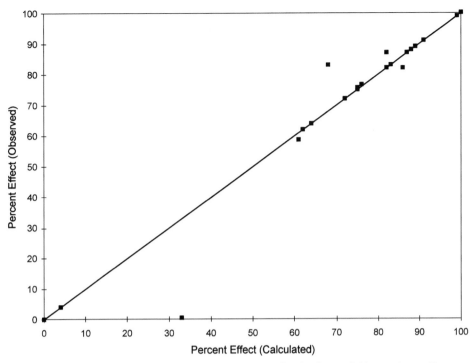

FIGURE 8.1   Plot of calculated versus observed percent effect for the set of chloromethanesulfon-ames against citrus red mites using neural network analysis based on E-State indices.

C. Concluding Remarks

The data set used in this study is relatively small for thorough neural network analysis. In fact, this set was too small to permit appropriate external validation. With this in mind, these results must be taken as indicative rather than final. For design of more effective sulfonamides, the chemist should direct attention to those atoms identified in the analysis, and attention should especially be directed to the effect of substituents on the sulfonamide oxygen atom and the amide nitrogen atom. Furthermore, the appearance of the long path-length chi index, $^{7}\chi_{P}{}^{v}$, indicates the importance of longer substituents. The implications of these findings are explored in more depth by Hall *et al.* (1998). Other data sets of larger size are under investigation using the methods of Parham. Preliminary results, such as those reported here, suggest that this approach is very useful when complete dose–response data are not available.

XII. SUMMARY

The examples in this chapter indicate the strength of the topological index approach to modeling properties. The use of whole molecule indices along with atom E-State indices is shown to be a powerful tool for modeling. Good models are obtained as indicated by sound statistics. Furthermore, interpretation in terms of molecular structure is possible because of the structure representation character of the indices.

XIII. A LOOK AHEAD

The utility of atom-type E-State indices has been demonstrated in this chapter. Some of the models used these newer structure representations. To provide a fuller exposition of the atom-type E-State indices and their power in QSAR modeling, Chapter 9 deals with models which use only the atom-type E-State.

REFERENCES

Arvidsson, L., Hacksell, U., and Glennon, R. (1986). Recent advances in central 5-hydroxytrypt-amine receptor agonists and antagonists. *Prog. Drug Res.* **30**, 365–471.

Bohmont, B. L., McCarty, R. H., Hortvedt, J. J., and Terry, D. L. (1987). *Pesticide Dictionary in Farm Chemical Handbook—'87*. Meister, Willoughby, OH.

Burnett, J. C. (1998). A QSAR model of the opiate receptor activity of various fentanyl-like compounds. Personal communication.

Dunmire, D., and Hall, L. H. (1998). Investigation of E-State, free valence and bond orders: Inhibition of MAO by phenethylamines. Personal communication.

Gough, J., and Hall, L. H. (1998a). QSAR analysis of carbiquinones as antileukemic agents: Combination of E-State indices and molecular connectivity indices. *J. Chem. Inf. Comput. Sci.* (in press).

Gough, J., and Hall, L. H. (1998b). Modeling the toxicity of amide herbicides using the electro-topological state. *Environ. Tox. Chem.* (in press).

Grella, B. S.(1998). A QSAR model for the affinities of various phenylisopropylamine derivatives at the 5-HT$_2$ binding sites. Personal communication.

Hall, L. H., and Kier, L. B. (1984). Molecular connectivity of phenols and their toxicity to fish. *Bull. Environ. Contam. Toxicol.* **32**, 354–362.

Hall, L. H., and Kier, L. B. (1986). Molecular connectivity and total response surface optimization. *J. Mol. Struct. (Theochem.)* **134**, 309–316.

Hall, L. M., Parham, M., and Hall, L. H. (1998). Direct modeling of percent effect biological data: Neural network analysis with the E-State and chi indices. Personal communication.

Hammett, L. P. (1940). *Physical Organic Chemistry*. McGraw-Hill, New York.

Hansch, C. (1969). A quantitative approach to biochemical structure–activity relationships. *Acc. Chem. Res.* **2**, 232–239.

Hasagawa, K., Kimura, T., Miyashita, Y., and Funatsu, K. (1996). Nonlinear partial least squares modeling of phenyl alkylamines with the monoamine oxidase inhibitory activities. *J. Chem. Inf. Comput. Sci.* **36**, 1025–1029.

Huuskonen, J., Salo, M., and Taskinen, J. (1997). Neural network modeling for estimation of the aqueous solubility of structurally related drugs. *J. Pharm. Sci.* **86**, 450–454.

Huuskonen, J., Salo, M., and Taskinen, J. (1998). Aqueous solubility prediction of drugs based on molecular topology and neural network modeling. *J. Chem. Inf. Comput. Sci.* **38**, 450–456.

Ickikawa, H., and Aoyama, T. (1991). How to see characteristics of structural parameters in QSAR analysis: Descriptor mapping using neural networks. *SAR QSAR Environ. Res.* **1**, 115–130.

Jager, G. (1983). Herbicides. In *Chemistry of Pesticides* (K. H. Buchel, Ed.; G. M. Holmwood, Trans.), pp. 322–392. Wiley, New York.

Kier, L. B. (1980). Structural information from the molecular connectivity $^4\chi_{pc}{}^v$ index. *J. Pharm. Sci.* **69**, 1034–1039.

Kier, L. B. (1987). Indexes of molecular shape from chemical graphs. *Med. Res. Rev.* **7**, 417–440.

Kier, L. B., and Hall, L. H. (1976a). *Molecular Connectivity in Chemistry and Drug Research*. Academic Press, New York.

Kier, L. B., and Hall, L. H. (1976b). Molecular connectivity VII: Specific treatment of heteroatoms. *J. Pharm. Sci.* **65**, 1806–1809.

Kier, L. B., and Hall, L. H. (1986). *Molecular Connectivity in Structure–Activity Analysis*. Wiley, New York.

Kier, L. B., and Hall, L. H. (1995). A QSAR model of the OH radical reaction with CFCs. *SAR QSAR Environ. Res.* **3**, 97–100.

Kier, L. B., Murray, W. J., Randic, M., and Hall, L. H. (1976). Molecular connectivity V: Connectivity series concept applied to density. *J. Pharm. Sci.* **65**, 1226–1230.

Klopman, G., Wang, S., and Balthasar, D. (1992). Estimation of aqueous solubility of organic molecules by group contribution approach. Application to the study of biodegradation. *J. Chem. Inf. Comput. Sci.* **32**, 474–482.

Lipinski, C., Lombardo, F., Dominy, B., and Feeney, P. (1997). Experimental and computational approach to estimate solubility and permeability in drug discovery and development settings. *Adv. Drug Delivery Rev.* **23**, 3–25.

Meylan, W., Howard, P., and Boethling, R. (1996). Improved method for estimating water solubility from octanol/water partition coefficients. *Environ. Toxicol. Chem.* **15**, 100–106.

Richard, A. J., and Kier, L. B. (1980). Structure–activity analysis of hydrogen MAO inhibitors using molecular connectivity. *J. Pharm. Sci.* **69**, 124–126.

Schaper, K. (1987). QSAR analysis of logit transformed dose response curves and dose response curve fragments of sets of compounds. In *QSAR in Drug Design and Toxicology* (D. Hadzy and B. Jerman-Blazic, Eds.), pp. 58–60. Elsevier, Amsterdam.

Schaper, K., and Saxena, A. (1991). QSAR analysis of dose and property dependent activity classes of homologous drugs. In *QSAR: Rational Approaches to the Design of Bioactive Compounds* (C. Silippo and A. Vittoria, Eds.), pp. 45–48. Elsevier, Amsterdam .

Schieck, C. L. (1998). A QSAR study of D_3 dopamine receptor ligands. Personal communication.

Stjernlof, P., *et al.* (1995). Structure–activity relationships in the 8-amino-6,7,8,9-tetrahydro-3H-benz[e]indole ring system. 1. Effects of substituents in the aromatic system on serotonin and dopamine receptor subtypes. *J. Med. Chem.* **69**, 2202–2216.

Taft, R. W. (1956). Separation of polar, steric, and resonance effects in reactivity. In *Steric Effects in Organic Chemistry* (M. S. Newman, Ed.), pp. 556–675. Wiley, New York.

Tamaru, A., Ogawa, H., Nishimuru, T., Takahashi, Y., and Sasaki, S. (1988). Structure–activity study of systemic acaricidal N-substituted chloromethanesulfonamides by the ORMUCS method. *J. Pesticide* **13**, 1–6.

Thomson, W. T. (1983). *Book II. Herbicides in Agricultural Chemicals,* Rev. ed., pp. 55–98. Thomson, Indianapolis, IN.

van Daele, P., DeBroyn, M. F. L., Boey, J. M., Sancsuk, S., Agten, J. T. M., and Janssen, P. A. J. (1995). Synthetic analgesics: N-91-[2-arylethyl]-4-substituted 4-piperidinyl) N-arylalkanamides. *Arzneim.-Forsch.* 26, 1521–1529.

Wiese, M. (1991). Application of neural networks in the analysis of percent effect biological data. *Quant. Structure–Activity Relat.* 10, 369–371.

Wiese, M., and Schaper, K. (1993). Application of neural networks in the QSAR analysis of percent effect biological data: Comparison with adaptive least squares and nonlinear regression analysis. *SAR QSAR Environ. Res.* 1, 137–152.

Yalkowsky, S. H., and Banerjee, S. (1992). *Aqueous Solubility. Methods of Estimation for Organic Compounds.* Dekker, New York.

Zakarya, D., Larfaoui, E. M., Boulaamail, A., and Lakhlifi, T. (1996). Analysis of structure relationships for a series of amide herbicides using statistical methods and neural networks. *SAR QSAR Environ. Res.* 5, 269–279.

CHAPTER 9

Structure–Activity Studies: Atom-Type Indices

I. USE OF ATOM-TYPE INDICES

The E-State indices are formulated and computed for each atom or hydride group in the molecular structure. In Chapter 7, several examples were given showing the application of these atom-level indices to quantitative structure–activity relationship (QSAR) analyses. Those applications require that there is some portion of the molecular skeleton which is common to every structure in the data set. This topological superposition approach yields valuable information about regions of importance in the molecular skeleton, leading to greater understanding of the relation between structure and activity. However, the advantage of this approach is also a limitation when the data set contains an array of different molecular skeletons. For this reason, we have introduced an extension of the E-State formalism, called the atom-type E-State, in which the atoms and hydride groups in the molecule are classified according a valence state scheme (see Chapter 4). The most commonly used atom-type E-State index is the sum of the E-State indices for a specific atom type. (In an alternative approach the average E-State value may be used for each atom type.) In this manner an index is computed for each atom type in the data set. These atom-type E-State indices serve well as the basis for structure–activity analysis. In this chapter, we present examples of this mode of analysis.

II. TOXICITY OF SUBSTITUTED BENZENES TO FATHEAD MINNOWS

The United States Environmental Protection Agency in Duluth, Minnesota, established a program for the generation of high-quality fish toxicity data. These data have been published in a series of volumes (Brooke *et al.*, 1984) and have provided the basis for several QSAR investigations (Hall and Kier, 1984, 1986; Hall *et al.*, 1985, 1989a,b). We report here the first application of the E-State method to these data—an approach which is more general than earlier methods.

We consider a data set which consists of 141 substituted benzenes (Table 9.1;

TABLE 9.1   Modeling Fathead Minnow Toxicity with Atom-Type E-State Indices[a]

No.	Compound name	pLD_{50}	Calcu-lated	Residual
1	Benzene	3.40	3.40	0.00
2	Bromobenzene	3.89	3.91	−0.02
3	Chlorobenzene	3.77	4.01	−0.24
4	Hydroxybenzene	3.51	3.19	0.32
5	1,3,5-Trichloro-2-hydroxybenzene	4.40	4.51	−0.11
6	1,2-Dichlorobenzene	4.30	4.52	−0.22
7	1,3-Dichlorobenzene	4.62	4.51	0.11
8	1,4-Dichlorobenzene	4.02	3.80	0.22
9	1-Chloro-2-hydroxybenzene	3.84	4.07	−0.23
10	1-Chloro-3-methylbenzene	4.33	4.06	0.27
11	1-Methyl-2,3,6-trinitrobenzene	3.04	3.08	−0.04
12	1-Chloro-4-methylbenzene	3.21	3.25	−0.04
13	1,3-Dihydroxybenzene	3.77	3.48	0.29
14	1-Hydroxy-3-methoxybenzene	3.29	3.50	−0.21
15	1-Aminobenzene	3.58	3.45	0.13
16	1-Hydroxy-2-methylbenzene	3.36	3.37	−0.01
17	1-Hydroxy-3-methylbenzene	3.07	3.40	−0.33
18	1-Hydroxy-4-methylbenzene	3.48	3.90	−0.42
19	1-Hydroxy-4-nitrobenzene	4.21	3.88	0.33
20	1,4-Dimethoxybenzene	3.57	3.59	−0.02
21	1-Amino-2-chlorobenzene	3.63	3.59	0.04
22	1,2-Dimethylbenzene	3.76	3.61	0.15
23	1,4-Dimethylbenzene	4.38	3.90	0.48
24	1-Methyl-2-nitrobenzene	3.48	3.67	−0.19
25	1-Methyl-3-nitrobenzene	3.24	3.66	−0.42
26	1-Aldehydo-2-nitro-5-hydroxybenzene	3.35	3.64	−0.29
27	1-Methyl-4-nitrobenzene	3.80	3.71	0.09
28	1,3-Dinitrobenzene	3.80	3.72	0.08
29	1-Amino-2-methyl-3-nitrobenzene	3.79	3.72	0.07
30	1-Amino-2-methyl-4-nitrobenzene	3.77	3.67	0.10
31	1-Amino-2-methyl-5-nitrobenzene	4.89	5.08	−0.19
32	1-Amino-2-methyl-6-nitrobenzene	5.00	5.06	−0.06
33	1-Aldehydo-3-methoxy-4-hydroxy-5-bromobenzene	4.74	5.04	−0.30
34	1-Amino-3-methyl-6-nitrobenzene	4.30	4.27	0.03
35	1-Amino-2-nitro-4-methylbenzene	4.74	4.47	0.27
36	1-Amino-3-nitro-4-methylbenzene	4.54	4.47	0.07
37	1,2,3-Trichlorobenzene	3.86	3.88	−0.02
38	1,2,4-Trichlorobenzene	3.75	3.88	−0.13
39	1-Aldehydo-2-methylbenzene	3.90	3.82	0.08
40	1,3,5-Trichlorobenzene	4.04	4.34	−0.30
41	1,3-Dichloro-4-hydroxybenzene	4.21	4.37	−0.16
42	1,2-Dichloro-4-methylbenzene	5.01	5.04	−0.03
43	1,3-Dichloro-4-methylbenzene	3.75	4.08	−0.33
44	1-Aldehydo-3-methoxy-4-hydroxybenzene	3.99	4.03	−0.04
45	1-Hydroxy-2,4-dimethylbenzene	5.08	5.03	0.05
46	1-Hydroxy-2,6-dimethylbenzene	3.91	4.06	−0.15
47	1-Hydroxy-3,4-dimethylbenzene	4.12	4.41	−0.29
48	1-Hydroxy-2,4-dinitrobenzene	5.34	5.33	0.01
49	1,2,4-Trimethylbenzene	4.26	4.35	−0.09

TABLE 9.1   (*Continued*)

No.	Compound name	pLD$_{50}$	Calcu-lated	Residual
50	1-Methyl-2,3-dinitrobenzene	4.21	4.52	−0.31
51	1-Aldehydo-2-hydroxy-3,5-dibromobenzene	4.18	4.68	−0.50
52	1-Methyl-2,4-dinitrobenzene	4.46	4.26	0.20
53	1-Methyl-2,6-dinitrobenzene	4.70	4.68	0.02
54	1-Methyl-3,4-dinitrobenzene	5.43	5.54	−0.11
55	1-Methyl-3,5-dinitrobenzene	5.85	5.53	0.32
56	1-Amino-2-methyl-3,5-dinitrobenzene	4.88	5.60	−0.72
57	1-Aldehydo-2-fluorobenzene	6.06	5.84	0.22
58	1-Amino-2-methyl-3,6-dinitrobenzene	3.56	3.59	−0.03
59	1-Amino-2,4-dinitro-3-methylbenzene	4.33	4.46	−0.13
60	1-Amino-2,6-dinitro-3-methylbenzene	4.07	4.18	−0.11
61	1-Amino-2,6-dinitro-4-methylbenzene	3.60	3.66	−0.06
62	1-Amino-3,5-dinitro-4-methylbenzene	3.93	4.11	−0.18
63	1,3,5-Tribromo-2-hydroxybenzene	4.74	5.00	−0.26
64	2-Allylphenol	6.09	6.25	−0.16
65	1,2,3,4-Tetrachlorobenzene	5.93	5.64	0.29
66	1,2,3,5-Tetrachlorobenzene	4.38	4.34	0.04
67	1-Methyl-2,4,6-trinitrobenzene	3.73	3.73	0.00
68	1-Hydroxy-2,3,4,5,6-pentachlorobenzene	4.00	4.04	−0.04
69	1-Amino-4-bromobenzene	3.42	3.44	−0.02
70	4-Butylphenol	4.72	4.52	0.20
71	1-Amino-3,4-dichlorobenzene	4.99	4.96	0.03
72	1-Amino-2,4-dinitrobenzene	4.81	4.64	0.17
73	1-Amino-2-chloro-4-methylbenzene	4.02	4.07	−0.05
74	1-Amino-2-chloro-4-nitrobenzene	4.14	4.33	−0.19
75	4-*t*-Butylphenol	3.92	3.87	0.05
76	1-Amino-2,3,4-trichlorobenzene	5.19	5.05	0.14
77	1,3,5-Trichloro-2,4-dinitrobenzene	5.31	5.33	−0.02
78	1-Amino-2,3,5,6-tetrachlorobenzene	4.73	4.76	−0.03
79	1-Cyano-3,5-dibromo-4-hydroxybenzene	4.02	3.79	0.23
80	1-Cyano-2-amino-5-chlorobenzene	4.81	3.80	1.01
81	1-Cyano-2-chloro-6-methylbenzene	4.83	4.81	0.02
82	1-Cyano-2-methylbenzene	3.69	3.71	−0.02
83	4-Pentylphenol	3.70	4.07	−0.37
84	1-Aldehydo-2-chloro-5-nitrobenzene	3.82	3.69	0.13
85	1-Aldehydo-2,4-dichlorobenzene	5.25	5.24	0.01
86	1-Aldehydo-4-chlorobenzene	4.23	4.65	−0.42
87	1-Aldehydo-2-nitrobenzene	4.56	4.52	0.04
88	1-Aldehydobenzene	4.21	4.37	−0.16
89	4-Nonylphenol	2.87	2.93	−0.06
90	1-Aldehydo-2,4-dimethoxybenzene	3.39	3.59	−0.20
91	1-Aldehydo-2-hydroxy-5-bromobenzene	3.79	3.57	0.22
92	1-Aldehydo-2-hydroxy-5-chlorobenzene	3.34	3.48	−0.14
93	1-Aldehydo-2-hydroxybenzene	3.48	3.46	0.02
94	1-Aldehydo-2-methoxy-4-hydroxybenzene	3.80	3.52	0.28
95	2,4-Dichlorophenol	3.77	3.64	0.13
96	1-Aldehydo-3-methoxy-4-hydroxybenzene	3.52	3.59	−0.07
97	1-Aldehydo-2-hydroxy-4,6-dimethoxybenzene	3.51	3.63	−0.12
98	1-Amino-2,3,4,5,6-pentafluorobenzene	4.39	4.24	0.15

TABLE 9.1   (*Continued*)

No.	Compound name	pLD$_{50}$	Calcu-lated	Residual
99	1-Fluoro-4-nitrobenzene	4.12	4.19	−0.07
100	2,3,4,5-Tetrachlorophenol	4.21	4.65	−0.44
101	1-Amino-4-fluorobenzene	4.26	4.20	0.06
102	1-Aldehydo-2,3,4,5,6-pentafluorobenzene	4.46	4.18	0.28
103	1-Aldehydo-2-chloro-6-fluorobenzene	4.92	4.56	0.36
104	1-Acyl-4-chloro-3-nitrobenzene	5.29	5.29	0.00
105	1-Acyl-2,4-dichlorobenzene	2.60	3.29	−0.69
106	1-Acylbenzene	2.87	2.88	−0.01
107	4-*t*-Pentylphenol	4.09	4.70	−0.61
108	3-Nitrobenzonitrile	5.00	4.94	0.06
109	4-Nitrobenzonitrile	4.16	4.40	−0.24
110	2-Amino-4-nitrotoluene	3.88	3.50	0.38
111	2-Amino-6-nitrotoluene	4.63	4.65	−0.02
112	3-Amino-4-nitrotoluene	4.91	4.76	0.15
113	2-Phenylphenol	3.32	3.55	−0.23
114	4-Amino-2-nitrotoluene	5.07	5.17	−0.10
115	3-Methyl-2-nitrophenol	5.07	5.00	0.07
116	5-Methyl-2-nitrophenol	3.65	3.75	−0.10
117	1,5-Dimethyl-2,4-dinitrobenzene	4.27	3.97	0.30
118	4-Phenylazophenol	5.15	4.64	0.51
119	2-Amino-4,6-dinitrotoluene	4.33	4.63	−0.30
120	3-Amino-2,4-dinitrotoluene	6.37	5.71	0.66
121	3-Amino-2,6-dinitrotoluene	3.56	3.52	0.04
122	4-Amino-2,6-dinitrotoluene	4.34	4.00	0.34
123	2,4-Dinitro-5-methylphenol	3.60	3.80	−0.20
124	1-Naphthol	3.59	3.95	−0.36
125	1,3,5-Trinitrobenzene	3.36	3.96	−0.60
126	2-Phenoxyethanol	3.26	3.86	−0.60
127	Acetophenone	5.52	5.63	−0.11
128	Benzophenone	4.96	4.37	0.59
129	2,3,4-Trichloroacetophenone	3.93	4.02	−0.09
130	2,4-Dichloroacetophenone	4.47	4.46	0.01
131	1-Chloronaphthalene	4.46	4.61	−0.15
132	2,6-Dimethoxytoluene	5.18	5.18	0.00
133	Diphenylether	6.20	6.18	0.02
134	*p*-Nitrophenylphenylether	4.30	4.18	0.12
135	5-Amino-2,4-dinitro-1-methylbenzene	5.72	5.26	0.46
136	Methylbenzene	4.82	4.82	0.00
137	1,2-Dinitrobenzene	4.45	5.09	−0.64
138	1,4-Dinitrobenzene	5.26	4.68	0.58
139	1-Amino-3-nitro-4-hydroxybenzene	4.53	4.15	0.38
140	1-Chloro-2-methyl-4-hydroxybenzene	4.85	4.75	0.10
141	1-Methyl-2,5-dinitrobenzene	4.91	4.45	0.46

[a] Compounds are not named by IUPAC rules but simply as benzene with substituents as was done to emphasize the substituents (Hall and Kier, 1984).

Fig. 9.1). The data were collected using the same standard procedures (Brooke
et al., 1984; Hall and Kier, 1984) to obtain the 96-hr flowthrough LC_{50} value
for toxicity on the fathead minnow (*Pimephales promelas*). The toxicity was
converted to a molar basis and then to the negative logarithm, pLC_{50}. There are
nine functional groups in the data set: amino, cyano, nitro, hydroxy, keto,
fluoro, chloro, bromo, and aldehydo. The atom-type E-State indices used in the
analysis include eight for type of carbon atom in addition to nine from the
substituents $S^T(-CH_3)$, $S^T(-CH_2-)$, $S^T(>CH-)$, $S^T(-CH=)$, $S^T(>C=)$,
$S^T(-C\equiv)$, $S^T(...CH...)$, $S^T(...C...)$, $S^T(-NH_2)$, $S^T(-N=)$, $S^T(-OH)$,
$S^T(=O)$, $S^T(-F)$, $S^T(-Cl)$, and $S^T(-Br)$. Furthermore, in our earlier investigations of these data, we developed two additional variables to represent special
cases: an indicator variable for the presence of two nitro groups either *ortho* or
para to each other and an indicator variable for the presence of an aldehyde
group *ortho* to an OH group (as in salicylaldehyde).

The 17 variables were entered into an artificial neural network analysis.

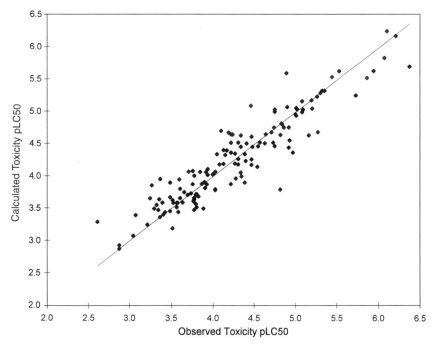

FIGURE 9.1   Plot of calculated toxicity (pLC_{50}) versus observed toxicity for 141 substituted benzenes against fathead minnow based on atom-type E-State indices in a 17:4:1 artificial neural
network.

After preliminary investigations we determined that a 17:4:1 network architecture appeared appropriate for this study. Fewer hidden neurons gave significantly larger errors and more hidden neurons gave little improvement. A test set of 23 compounds (16.3%) was set aside on a random basis and the network trained on the remaining 118 compounds (as a training set). The learning rate was set at 0.7 and the momentum at 0.5. The final analysis was run for 10,000 cycles. All network analysis were performed with the Neuralyst software (Cheshire Engineering Corp., 650 Sierra Madre Villa Avenue, Pasadena, CA 91107).

The final statistics indicate that the trained network provides a reliable basis for the model and for prediction, based on the correlation coefficient and mean absolute error (MAE):

Training set: $r^2 = 0.86$, MAE = 0.19; Test set: $r^2 = 0.89$, MAE = 0.24

When calculated pLC_{50} values are plotted against the observed values, the slopes and intercepts are as follows:

Training set: slope = 1.00, intercept = -0.01; Test set: slope = 1.01, intercept = -0.13

As these values suggest, there is a close correspondence between observed and calculated pLC_{50} values which is shown in Fig. 9.1. A plot of the residuals versus the observable toxicity shows no trends and appears to be random. There are five residuals in the training set (5.4%) and only one in the test set (4.3%) which are ≥ 2 SD. This distribution of residuals is quite acceptable. The two largest residuals in the training set are 1.01 for compound 80 (1-aldehydo-3-methoxy-4-hydroxybenzene) and 0.72 for compound 56 (1-methyl-2,4,6-trinitrobenzene). In the test set the largest residual is 0.61 for compound 118 (4-*tert*-pentylphenol).

These results indicate a reasonable model, especially for such a wide range in toxicity values [3.5 orders of magnitude (2.87–6.37)] and for such a diverse set of substituents (nine substituents and eight additional carbon atom types). Furthermore, the compounds at the extreme toxicity values tend to be well predicted. Most of these residuals are near the MAE value. Compound 120 (3-amino-2,4-dinitrotoluene) with $pLC_{50} = 6.37$ has a residual of 0.66. In general, the residuals on both training and test sets are quite acceptable.

III. BOILING POINT AND CRITICAL TEMPERATURE OF A SET OF HETEROGENEOUS COMPOUNDS

Physicochemical properties are often investigated with QSAR methods because of the availability of larger data sets which cover a wide range of skeletal structure. Boiling point and critical temperature are significant properties in revealing the intermolecular forces of molecules. Furthermore, they are useful for testing development of QSAR models. For both these reasons, we examined the relationship of boiling point and critical temperature to the atom-type E-State indices.

A set of data was selected from the DIPPR database by Egolf *et al.* (1994) for the purpose of testing a particular approach to QSAR modeling. We have chosen to use this same data set for modeling with the atom-type E-State indices both as a test of their QSAR capability and as a means of providing a model for prediction of boiling point and critical temperature. It is expected that the relation between these temperature measures and at least some of the molecular descriptors is nonlinear. For this reason, we elected to use artificial neural networks for modeling.

A. BOILING POINT

The descriptors used in this study to represent the molecular structure of this diverse data set are the atom-type E-State indices. The symbols for the 19 atom types represented in this data set are as follows: $S^T(-CH_3)$, $S^T(-CH_2-)$, $S^T(>CH-)$, $S^T(>C<)$, $S^T(...CH...)$, $S^T(...C...)$, $S^T(=CH_2)$, $S^T(-CH=)$, $S^T(>C=)$, $S^T(-C\equiv)$, $S^T(-NH_2)$, $S^T(-NH-)$, $S^T(>N-)$, $S^T(...N...)$, $S^T(-OH)$, $S^T(-O-)$, $S^T(=O)$, $S(-Cl)$, and $S(-Br)$. Structure input for the 298 compounds was from SMILES code. Thirty compounds were selected as the testing set. These results have been reported by Hall and Story (1996).

The neural network analysis was begun by considering several network architectures in order to find a useful model. We examined only three-layer networks with 19 input nodes (for the 19 structure variables), one output node for the boiling point, and one hidden layer with n nodes. For these $19:n:1$ networks, we found that the root mean square error improved significantly for n up to values of approximately 4 or 5. A $19:5:1$ network was selected for development of a boiling point model. In these studies, the learning rate was set at 0.80 and the momentum at 0.60. The fraction of the data set selected for training was approximately 90%. For each training set, we examined several sets of randomly selected weights. Typical runs lasted 20,000–50,000 cycles.

For the several runs investigated the overall relative error (for the whole data set of 298 compounds) ranged from 0.94% to 1.17%. The results for the best run are given in Table 9.2. The MAEs for the whole set, training set, and testing set are 3.93, 3.86, and 4.57 K, respectively, for the best run. We also ran and averaged 10 runs; the average relative percentage error for these runs is 0.94% for the total set and 1.12% for the test set. The plot of calculated versus observed boiling point is given in Fig. 9.2. The complete tables of data are reported by Hall and Story (1996) and are not provided here.

TABLE 9.2   Statistical Summary for Boiling Point of Mixed Organic Compounds with Atom-Type E-State Indices and an Artificial Neural Network

	Training set	Testing set	Total set
No. of compounds	268	30	298
Mean absolute error	3.86	4.57	3.93
Root mean square error	5.30	5.87	5.36
Correlation coefficient, r (calculated vs predicted)	0.998	0.997	0.998
Slope of line (calculated vs predicted)	1.00	0.98	1.00

The molecules in this set are characterized by 19 atom types (listed earlier), including five elements: carbon, nitrogen, oxygen, chlorine, and bromine. In addition to single, double, triple, and aromatic bonds, the following functional groups are present: chloro, bromo, alcohol (including diols), cyano, amino, ether, carboxylic acid, ester, and carbonyl. Nonpolar and polar molecules are included as well as those exhibiting hydrogen bonding. The boiling points range from 225 K (propylene) to 648 K (stearic acid).

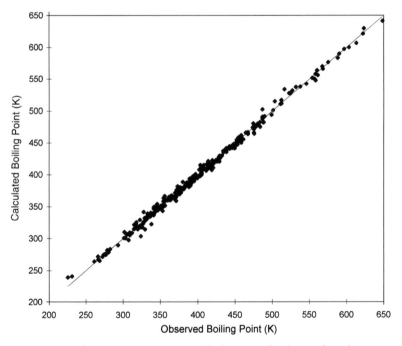

FIGURE 9.2   Plot of calculated versus observed boiling point for the set of 298 heterogeneous organic compounds based on a 19:5:1 artificial neural network using atom-type E-State indices.

For purposes of comparison, a linear model was obtained using multiple linear regression. For the same 19 input variables as used in the neural network, the standard error was found to be much larger (15.6 K). Furthermore, there were many more large residuals: 40 residuals were larger than 20 K and 5 were larger than 40 K. Thus, it appears that the nonlinear nature of the neural network model is a better vehicle for modeling boiling point.

The results of this model are excellent. The mean absolute error for the whole data set and for the test set is under 4 K. This value is the lowest MAE for such a large, heterogeneous data set published to date. The average relative error for the whole data set is 1.05% and it is 1.12% for the test set. These values compare very favorably with the estimated experimental errors (Egolf et al., 1994). The slope of the line in the plot of calculated versus observed (Fig. 9.2) is 1.00 for the whole set and 0.98 for the test set; this is a very gratifying result. The value of unity for the slope indicates that over the range of input variables, there is a strong trend between calculated and observed boiling point.

This study reveals that for the boiling point, the electronic structure information of the E-State is well related to the intermolecular forces present. Three types of intermolecular forces occur in these molecules: dispersion, dipolar, and hydrogen bonding. In addition, there are several structure types present in this heterogeneous data set. The high quality of the QSAR model for boiling point clearly indicates the strong modeling ability of the atom-type E-State indices and confirms that the E-State is a valuable representation of molecular structure.

B. CRITICAL TEMPERATURE

For investigation of critical temperature the set of 165 compounds and their critical temperatures (K) were obtained from Egolf et al. (1994). Eighteen compounds comprised the testing set. For critical temperature, the values range from 365 K (propylene) to 782 K (quinoline). The neural network analysis was done in a manner similar to that for the boiling point described in the previous section.

A 19:4:1 architecture was determined to be adequate for the neural network analysis (Fig. 9.3). The overall relative error for the best run (for the whole data set of 165 compounds) ranged from 0.97 to 1.17%. The MAEs for the whole set, training set, and testing set are 4.52, 4.39, and 5.59 K, respectively, for the best run. Average relative percentage error for five runs is 0.77% for the total set and 0.95% for the test set. The summary for the selected model is given in Table 9.3. A plot of calculated versus observed boiling point is given in Fig. 9.2. The complete table of compounds and their observed, calculated, and residual boiling points are given by Hall and Story (1996).

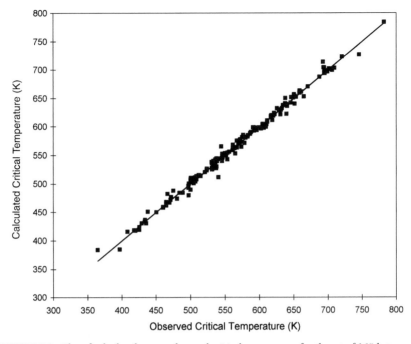

FIGURE 9.3   Plot of calculated versus observed critical temperature for the set of 165 heterogeneous organic compounds based on a 19:4:1 artificial neural network using atom-type E-State indices.

For this critical temperature model, only three compounds have residuals ≥20 K: *n*-nonadecane, 2,2,4-trimethylpentane, and *n*-octadecane. Sixteen residuals lie between 10 and 20 K. In the test set, only two compounds have residuals ≥10 K; none have residuals ≥12 K. The plot of calculated versus observed critical temperature (Fig. 9.2) represents an excellent relationship, with the points located very close to the line. The plot of the residuals versus

TABLE 9.3   Statistical Summary for Critical Temperature of Mixed Organic Compounds with Atom-Type E-State Indices and an Artificial Neural Network

	Training set	Testing set	Total set
No. of compounds	147	18	165
Mean absolute error	4.39	5.59	4.52
Root mean square error	6.61	6.71	6.62
Correlation coefficient, r (calculated vs predicted)	0.997	0.995	0.996
Slope of line (calculated vs predicted)	1.00	1.00	1.00

observed critical temperature (not shown) appears to be random, showing no pattern. There seems to be no particular pattern among the types of compounds found with the residuals ≥ 10 K. When the calculated critical temperature values are compared to the observed values, the plot yields a slope of 1.00 for the test set as well as for the training set. These results show that, over the range of the input values, there is a strong relationship between the calculated and observed values.

The summary results for this model are excellent. The MAEs are 4.5 and 5.6 K for the whole data set and the test set, respectively. The average relative error is 0.94% for the whole set and 0.97% for the test set, which compares favorably with the estimated experimental errors (Egolf et al., 1994). These averages are excellent for these data, which have not been widely modeled. For comparison, Egolf et al. (who used fewer variables) found a standard deviation of 11.88 K for the training data set (147 compounds) compared to 7.1 K found in the atom-type E-State model (modeling results were not given for the test set by Egolf et al.).

For purposes of comparison, a linear model was obtained using multiple linear regression. For the same 19 input variables as used in the neural network, the standard error was found to be much larger (23.7 K). Furthermore, there were many more large residuals, including 49 larger than 20 K and 8 larger than 40 K. Thus, it appears that the nonlinear nature of the neural network model makes it is a better vehicle for modeling boiling point. Furthermore, the improvement over the linear model is even greater for critical temperature than for boiling point.

C. Concluding Remarks on Boiling Point and Critical Temperature

For properties such as the boiling point and critical temperature, the electronic structure information of the E-State is related to the intermolecular forces present. For this data set, three types of intermolecular forces are considered present: dispersion, dipolar, and hydrogen bonding. Furthermore, there are many structure types present in this heterogeneous data set, related to more than a dozen functional groups. The high quality of the QSAR models for both boiling point and critical temperature clearly demonstrates the strong modeling ability of the atom-type E-State indices and confirms that the E-State is a powerful representation of molecular structure. The artificial neural network models developed here also seem to be very useful in modeling properties such as boiling point and critical temperature. We are continuing to explore the usefulness of artificial neural networks for modeling and the atom-type E-State indices as structure descriptors for heterogeneous data sets on physicochemical properties and for biological data sets.

IV. BOILING POINT OF A SET OF 372 POLYCHLOROALKANES, ALCOHOLS, AND ALKANES

Boiling point is an important property for consideration in certain environmental problems, and it is a useful property for testing the development of QSAR models. For both these reasons, we have chosen to examine the relationship between boiling point and the atom-type E-State indices.

The volatility of many classes of compounds is important in environmental issues. Among these classes are the alkanes, alcohols, and chloroalkanes. We have created a large data set for modeling with the atom-type E-State indices both as a test of their QSAR capability and as a means of providing a model for boiling point prediction. It is expected that the relation between boiling point and at least some of the molecular descriptors is nonlinear. For this reason, we have elected to use artificial neural networks for modeling.

The descriptors used in this study to represent the molecular structure of the alkanes, alcohols, and chloroalkanes are the atom-type E-State indices. The symbols for the six atom types represented in this data set are $S^T(-CH_3)$, $S^T(-CH_2-)$, $S^T(>CH-)$, $S^T(>C<)$, $S^T(-OH)$, and $S^T(-Cl)$. Structure input for the 372 compounds was from SMILES code. This work was reported by Hall and Story (1997).

The boiling points of the alkanes were taken from Dreisbach (1959) for the complete set of alkanes isomers of butane through all the decane isomers in addition to normal alkanes up to 17 carbon atoms for a total of 154 compounds. Data for 108 saturated alcohols and 110 mono- and polychloroalkanes were taken from the *CRC Handbook* (Lide, 1994). The alcohols include methanol through eight undecanols and dodecanols, and the chloroalkanes include the chloromethanes and several mono- and dichlorohexanes and heptanes. The boiling point data were converted to the Kelvin scale. The list of compounds and their boiling points, together with the atom-type E-State indices for each compound, are given by Hall and Story (1997).

For the neural network analysis the boiling points and the six atom-type E-State indices were entered into the spreadsheet of the Neuralyst program, a modification of the EXCEL spreadsheet software. We used the sigmoid transfer function in our investigation and the back-propagation training algorithm. All the training runs were done interactively. A training set was selected randomly by an automatic procedure in the software using a preselected percentage of the data set.

Several network architectures were examined in order to find a useful model. We examined only three-layer networks with six input nodes (for the six structure variables), one output node (for the boiling point), and one hidden layer with n nodes. For these $6:n:1$ networks, we found that the RMS error improved significantly up to values of approximately 6 or 7 for n. We decided to further develop a $6:7:1$ network. In these studies, the learning rate was set at 0.80 and the momentum at 0.60. These values are close to those suggested in the Neuralyst software but we obtained these specific values through trial and error. The fraction of the data set selected for training was approximately 80%. Typical runs lasted 20,000–30,000 cycles.

Several runs were made as follows. Neuralyst was used to select a random set of compounds as the test set. Several runs were made, each with a different set of randomly selected starting weights. For the studies performed in this investigation the overall relative error (for the whole data set of 372 compounds) ranged from 0.97 to 1.17%. The MAE for the training set ranged from 3.99 to 4.90 and for the test set from 4.03 to 7.00. The results from the run which gave the best statistics were used as the model. The statistical summary for the selected model is given in Table 9.4. [The results of the selected model are given in Hall and Story (1997), including the compound name, the observed boiling point, values for the atom-type E-State indices, and the residual.] The plot of observed versus calculated boiling point is given in Fig. 9.4.

The results of the neural network modeling with the atom-type E-State indices are excellent as viewed from several perspectives. The overall relative error, 0.97%, appears to be excellent for this mixed set of data. For the purposes of prediction of boiling points for use in environmental work, these predictions are very useful because of the 1% error. Furthermore, we are unaware of literature reports of models which give an error on boiling point this low. In addition, this data set combines molecules with different types of intermolecular forces. The alkanes are nonpolar and characterized only by dispersion forces. The alcohols are polar and also characterized by hydrogen bonding in addition to dipolar intermolecular forces. The chloroalkanes are polar but exhibit no hydrogen bonding. These three types of molecules appear to be well treated.

Furthermore, in the data set there are primary, secondary, and tertiary alcohols. An examination of the residuals indicates that all three types are equally well treated. The $S^T(\text{—OH})$ index reflects the type of alcohol group; the $S^T(\text{—OH})$ index increases in the order, primary $<$ secondary $<$ tertiary, which is generally the inverse of the boiling point order. The $S^T(\text{—OH})$ index is significantly better than an indicator variable which, in this case, would be bivariate—

TABLE 9.4 Statistical Summary for Boiling Point of 154 Alkanes, 108 Alcohols, and 110 Chloroalkanes Using Atom-Type E-State Indices with an Artificial Neural Network

	Training set	Testing set	Total set
No. of compounds	294	78	372
Mean absolute error	3.99	4.03	4.00
Root mean square error	5.31	5.23	5.41
Slope of line for observed vs calculated boiling point	0.984	0.972	0.982

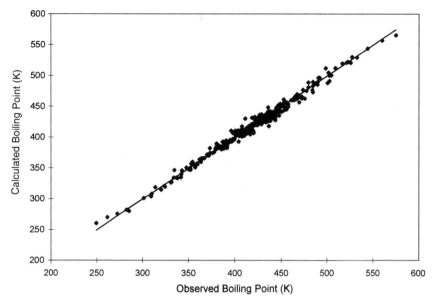

FIGURE 9.4 Plot of calculated boiling point (K) versus observed boiling point for 372 alkanes, alcohols, and chloroalkanes using a 6:7:1 artificial neural network with atom-type E-State indices.

nonzero for the alcohols and zero for the others. $S^T(-OH)$ is zero for alkanes and chloroalkanes, but it varies with structure variation for the alcohols. Skeletal branching also influences boiling point; these effects are described directly in the carbon atom-type E-State indices. In general, $S^T(-CH_3) > S^T(-CH_2-)$ $> S^T(>CH-) > S^T(>C<)$.

The predictive quality of this model is clearly indicated by the fact that the MAE for the test set (4.03 K) is essentially the same as that for the training set (4.00 K). In the test set, 29 are alkanes. Only 3 of these show residuals ≥ 10 K, with the largest being 11.8 K. Among the 26 alcohols in the test set, only 2 have residuals ≥ 10 K, with the largest being 10.8 K. Finally, among the 23 chloroalkanes in the test set, only 4 have residuals ≥ 10 K, with 11.3 K being the largest. In the total test set, 65.4% have residuals < 5 K.

For properties such as the boiling point, the electronic structure information is related to the intermolecular forces present. For this data set, three types of intermolecular forces are considered present: dispersion, dipolar, and hydrogen bonding. Based on the high quality of the QSAR model for boiling point shown here, the atom-type E-State indices appear to be a useful representation of molecular structure. We are continuing to explore the usefulness of these structure descriptors for more heterogeneous data sets on boiling point and for biological data sets.

V. SUMMARY

These QSAR analyses demonstrate the power of the atom-type E-State indices for heterogeneous sets of data. Excellent results are obtained and validated with testing sets. For these studies artificial neural networks have been used. The results are clearly superior to those obtained with multiple linear regression methods. Such results appear to support the basic concept that most QSAR relationships are actually nonlinear. Whether the method of using artificial neural networks will supplant regression methods remains to be seen. The power and utility of the atom-type E-State indices, however, seems clear.

This chapter illustrates a major advantage of the E-State methodology; that is, its ability to provide models for physicochemical properties and toxicity in addition to drug potency. The impact of this comprehensive structure representation is that in an overall drug design project, the same basic methodology can be used for a wide variety of associated problems. The same set of structure descriptors can be used to develop models and the same analysis can be used for structure interpretation.

VI. A LOOK AHEAD

The first nine chapters of this book have set forth the E-State formalism and demonstrated applications of the methodology in several ways. These chapters served to demonstrate the power of the E-State formalism. In Chapter 10, possible extensions of the basic approach are discussed. We encourage readers to think about further development of the E-State concepts.

REFERENCES

Brooke, L., Call, D., Geiger, D., and Northcott, C. (Eds.) (1984). *Acute Toxicities of Organic Chemicals to Fathead Minnows (Pimephales promelas).* Center for Lake Superior Environmental Studies, University of Wisconsin-Superior, Superior, WI.

Dreisbach, R. R. (Ed.) (1959). *Physical Properties of Compounds—II,* Advances in Chemistry Series No. 22. American Chemical Society, Washington, DC.

Egolf, L. M., Wessel, M. D., and Jurs, P. C. (1994). Prediction of boiling points and critical temperatures of industrially important organic compounds from molecular structure. *J. Chem. Inf. Comput. Sci.* 34, 947–956.

Hall, L. H., and Kier, L. B. (1984). Molecular connectivity of phenols and their toxicity to fish. *Bull. Environ. Contam. Toxicol.* 32, 354–362.

Hall, L. H., and Kier, L. B. (1986). SAR studies on the toxicities of benzene derivatives: II. An analysis of benzene substituent effects on toxicity. *Environ. Toxicol. Chem.* 5, 333–337.

Hall, L. H., and Story, C. T. (1996). Boiling point and critical temperature of a heterogeneous data set: QSAR with atom type electrotopological state indices using artificial neural networks. *J. Chem. Inf. Comput. Sci.* 36, 1004–1014.

Hall, L. H., and Story, C. T. (1997). Boiling point of a set of alkanes, alcohols and chloroalkanes: QSAR with atom type electrotopological state indices using artificial neural networks. *SAR QSAR Environ. Res.* **6**, 139–161.

Hall, L. H., Kier, L. B., and Phipps, G. (1985). SAR studies on the toxicities of benzene derivatives I. An additivity model. *J. Environ. Toxicol. Chem.* **3**, 333–337.

Hall, L. H., Maynard, E. L., and Kier, L. B. (1989a). Structure–activity relationship studies on the toxicity of benzene derivatives: III. Predictions and extension to new substituents. *J. Environ. Toxicol. Chem.* **8**, 431–436.

Hall, L. H., Maynard, E. L., and Kier, L. B. (1989b). QSAR investigation of benzene toxicity to fathead minnow using molecular connectivity. *J. Environ. Toxicol. Chem.* **8**, 783.

Lide, D. R. (Ed.) (1994). *CRC Handbook of Chemistry and Physics,* 75th ed. CRC Press, Boca Raton, FL.

CHAPTER 10

Future Directions of E-State Studies

No story in the field of rational drug design is ever complete. A decade or two from now we will look back on the prominent ideas and methods of today and reflect on their contributions and/or shortcomings. We will consider the manner in which they have pointed the way or have led to dead ends. The most important contribution that anyone can make in a very dynamic branch of applied science is to generate new ideas and concepts that will rise or fall with their intrinsic quality and the niches that they fill. The concept of the E-State is, we hope, such a contribution. It has and we believe it will continue to fill a niche in rational drug design as well as in structure modeling and prediction.

It is important to us that we present in this book a clear and lucid description and account of the utility of this method. It is also important that we devote space to speculation on possible improvements and derivations of the E-State method. This is a practice we have used in our earlier books on structure–activity methods. Therefore, we describe here some ideas in embryonic states in the hope that they will earn their place in the pantheon of nonempirical structure descriptors.

I. GROUP E-STATE INDICES

In a review of the E-State method, we proposed the possibility for calculating an approximate value of an E-State index for a collection of atoms in a group (Kier and Hall, 1993). For example, consider a methoxyl group (Scheme I).

$$--O--CH_3 \tag{I}$$

The standard practice is to consider the two atoms, $--O--$ and $--CH_3$, with their intrinsic states and to compute their E-State values in the context of the molecules in which they reside. For this calculation, the intrinsic state values are 3.50 for $I(--O--)$ and 2.00 for $I(--CH_3)$. A convenient consolidation of these two atoms into a single group with an appropriate intrinsic state would be a way to estimate the group's influence on nearby atoms. It would also permit a rapid (hand calculator) estimate of the influence of this group compared to that of other common groups.

One approach is to average the I values for —O— and CH_3 to yield a methoxy group I value of 3.288. Upon reflection, however, the oxygen atom of the group is always closer than the methyl group to any other atom in the molecule with its consequent influence. To account for this positional effect, the I values of each part of the group could be weighted according to their distance from other atoms in the molecule. An alternative is to find a value for a group intrinsic state that reproduced the normally calculated influence on an attached atom.

Consider a molecule (Scheme II). If the intrinsic state values of X are varied, for example, $I(X) = 2.0$, 4.0, and 6.0, the calculations of the E-States of X give the values shown in Table 10.1.

$$X\text{---}O\text{---}CH_3 \qquad\qquad (II)$$

If a group intrinsic state value for the methoxyl group is to be used, $I(OCH_3)$, it must have the values shown in Table 10.1 to reproduce the same E-State values calculated for atom X. It is obvious that the $S(X)$ value depends on $I(X)$ if a constant value of $I(OCH_3)$ is used. An average value of $I(OCH_3)$ might be used if a comparison is being made between the influence of this group and that of other groups on a common atom of the same graph distance. In addition, some modification would have to be made in $I(OCH_3)$ in order to reasonably reproduce normal results. The advantage to such a strategy must be weighed against the ease with which calculations can be made, even with a pocket calculator. The concept of a group intrinsic state value is useful in the situation in which comparisons are needed among groups. This situation is frequently en-

TABLE 10.1   Group Intrinsic State Values
Reproducing Calculated E-State Indices

$$X\text{---}O\text{---}CH_3$$

$I(X)$	Calculated $S(X)$	$I(OCH_3)$ to reproduce $S(X)$
2.0	1.625	3.500
4.0	4.347	2.612
6.0	7.069	1.724

countered with decision trees or substituent effect considerations. The concept
is worth further exploration.

II. AN ITERATIVE E-STATE

We have considered the possibility of exploring an iterative procedure to calculate indices in a molecule. In the spirit of Sanderson's (1976) electronegativity equalization, one bonded atom will diminish the electronegativity of the
other, thus ultimately producing a constant or weighted average value for each
atom. The E-State formalism provides for a single calculation of the influence
of two atoms bound directly or remotely to each other in a molecule:

$$\Delta I_A = \frac{I_A - I_B}{r_{AB}{}^2} \tag{10.1}$$

If $I_A > I_B$ then a positive ΔI_A will result and the E-State value will be greater
than the intrinsic state value: $S_A > I_A$. The reverse is true if $I_B > I_A$. If atom A is
enriched by drawing electron density to it from B, then I_A should decrease with
this encounter. The expectation might be a decrease in the $(\delta^v - \delta)$ electronegativity term of the I index (Eq. 10.2) if I is expressed in the full form described
in Chapter 2:

$$I = \frac{(\delta^v - \delta) + 1}{\delta + 1} \tag{10.2}$$

From this reasoning, the value of S_A may either be enhanced or diminished
by the contribution from I_B. This consideration is compounded by the presence
of the topological state index (δ) in the denominator of Eq. (10.2). This attribute is not influenced directly by electronegativity effects. It is a constant—a
statement of the architecture of the atom and attached neighbors.

In one possible approach, in the first cycle of iteration, a fractional multiplier
may be applied to the perturbation term ΔI_{ij}, e.g., $0.75\ \Delta I_{ij}$. All perturbations
are computed and applied to achieve the first-pass values for all $S(i)$. Then the
I values of each atom are modified by adding $-\Sigma \Delta I_{ij}$ to each I value. In this
fashion, the I values of the atoms with the larger I values are diminished by the
amount $0.75\ \Delta I_{ij}$. Correspondingly, the smaller I values are increased by the
same amount. In the second cycle of iteration the second-order perturbations
are computed from the modified I values, producing additional but smaller perturbations. This iterative process is continued until either the perturbations are
smaller than a preset limit or a preset number of cycles is reached. A similar
strategy could be pursued when only the electronegativity portion of the I value
is modified.

III. HIGHER DIMENSIONAL E-STATE CALCULATIONS

E-State values as currently calculated are based on the dimensionless molecular graph of a molecule; the graph is a statement of objects and their relations rather than a Cartesian representation of the structure. The "distances" are recorded as counts of atoms in the shortest paths separating atoms. As a consequence, the across-space influence and the geometric isomerism encountered in a molecule are not reflected in the calculation of the perturbation term ΔI_{ij}. Some exploratory work has been carried out and is worth describing here to stimulate progress.

A. CIS–TRANS ISOMERS

The consequences of *cis–trans* isomerism are not observed in the calculation of E-State indices because the graphs of both isomers show the same result. In order to enrich the information content of the state values, we suggest a possible extension to consider geometry to be used in E-State calculations. Consider the molecule pentene-2 in Fig. 10.1a. It can exist in either *cis* or *trans*

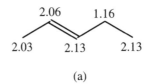

(a)

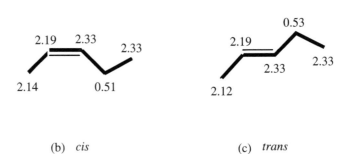

(b) *cis*   (c) *trans*

FIGURE 10.1   (a) The structure of pentene-2 with the E-State indices shown. (b) *cis* pentene-2 and (c) *trans* pentene-2, showing E-State values based on approximate Euclidean distances.

geometry, with each separated by a large energy barrier so that they are separable molecules with different properties. The E-State values, calculated as described earlier, are assigned to the atoms in Fig. 10.1a. The usual algorithm is

$$S_i = I_i + \Sigma_j \, \Delta I_{ij} \tag{10.3}$$

where ΔI_{ij} is expressed as in Eq. (10.1).

The value of r_{ij} is the count of atoms in the shortest path separating atom j from atom i. In this case, the geometry-based value of r_{ij} for the path separating the two methyl groups is 5. We then consider modifying this calculation using Euclidean distances based on the preferred conformation of the molecule in both its *cis* and *trans* geometries. The S value is now derived from Eq. (10.3), where ΔI_{ij} is expressed as in Eq. (10.4):

$$\Delta I_{ij} = \frac{\Sigma_j \, (I_i - I_j)}{(\text{Euclidean distance in angstrom})^2} \tag{10.4}$$

Figs. 10.1b and 10.1c show the pentene-2 molecule in both the *cis* and *trans* forms along with the geometry-based E-State values. The calculations are based on a planar conformation for the carbon skeleton for both *cis* and *trans* forms. The structure in Fig. 10.1a is labeled with the E-State values derived from the premise that there is a decline in influence from distant atoms as a function of the square of the path count separating those atoms. Clearly, geometric isomerism is not encoded. We now examine the effect of remote atoms on the intrinsic state values by adopting the hypothesis that electrotopological influences operate as a function of spatial distance irrespective of the formal bond structure. Accordingly, the influence is diminished as a function of Euclidean distance between any two atoms. These results are shown in Figs. 10.1b and 10.1c.

Note that we cannot directly compare the results in Fig. 10.1a with those in Figs. 10.1b and 10.1c because the two metrics used in defining the distance in the basic algorithm are different. The results, however, can be compared in relative terms. It is apparent that the trends of E-State values are the same for the two methods. Looking at the results in more detail, it is apparent that there are very modest differences between the *cis* and the *trans* E-State values. This is a very simple model with virtually no conformational uncertainty. The basic assumption here is that all interactive effects are across space and not along the axes of bonds. This cannot be totally correct, or even largely correct. Many atom pairs in a calculation such as this will have distance vectors paralleling bonds and therefore cross-space effects may be indistinguishable from through-bond effects.

B. CONFORMERS

The following question was posed in the previous section: How much of the electrotopological influence between any two atoms travels across space and how much travels through contiguous bonds? Recognizing some contribution from both, it may be possible to address the problem of reflecting this duality in the E-State value calculation. Consider the ethanolamine molecule in Fig. 10.2. The probable stable conformers are the *trans* and the *gauche* forms. The separation distances between the nonhydrogen atoms are shown in Figs. 10.2b and 10.2c. The Euclidean distances may be converted into the same metric as the graph bond distances by dividing them by a constant such as 1.54, the intercarbon distance in ethane. Euclidean distances can thus be expressed in a graph metric as seen in Fig. 10.2. For the calculation of E-State values, it can be assumed in this case that the influences through graph bonds and Euclidean space, expressed as equivalent bonds, are the same. The expression for the E-State value at atom i is

$$S_i = I_i + \Sigma_j \, \Delta I_{ij}, \text{ where } \Delta I_{ij} = \frac{I_i - I_j}{(r_g + r_E)^2} \qquad (10.5)$$

where r_g is the graph bond distance and r_E is the Euclidean distance in graph bond metric. Henceforth, we will refer to r_E as the relative Euclidean distance. Note that in these calculations, r_g is the count of bonds in the path plus 1, or the count of atoms in the path. To put r_E on the same scale, it is necessary to add 1 to the relative Euclidean distance in this expression. In summary,

$$r_g = \text{bonds in path} + 1 \qquad (10.6)$$

$$r_E = \text{relative Euclidean distance} + 1 \qquad (10.7)$$

The calculation results are shown in Fig. 10.2 expressed as E-State values at each atom. In this case the model of the effect of path distance and Euclidean distance is assumed to have an equal share of the influence on an atom, reflected by its E-State value. By converting Euclidean distance to equivalent bond distances, it is possible to integrate the two metrics into a single expression. Certainly, alternative schemes exist in which r_g and r_E may be combined in some additive manner with appropriate weighting. The proportion of r_g and r_E remains to be determined since this has not been the subject of study. It is tempting to resort to conventional intuition and say that the graph bond influence should be much greater. This may not be valid when two groups are near across space. The values will be approximately comparable, however, in many circumstances; the question of predominance is moot at this time.

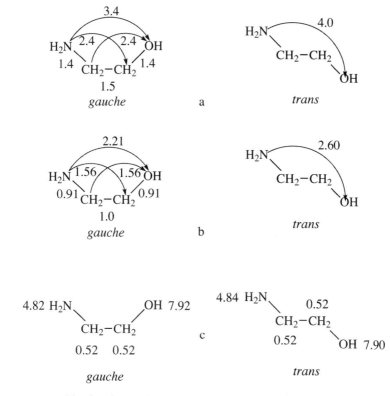

FIGURE 10.2   (a) Ethanolamine shown in both *gauche* and *trans* forms with the corresponding atom–atom distances in angstroms. (b) Ethanolamine shown in both *gauche* and *trans* forms with the atom–atom distances scaled to 1.5 Å. (c) E-State values computed with the scaled distances.

We have created a model in which the graph bond distances and the Euclidean distances among atoms in a molecule participate in the calculation of the E-State values for each atom. The method is sensitive to the assumed or calculated conformation. We provisionally adopt an equivalent role for these two dimensions. The method as described distinguishes among conformers giving E-State values as a function of conformational differences. Computer-based calculations using molecular or quantum mechanics should efficiently reveal conformational structure. The method is embryonic but it has a clear connection to experience. It is rapid and logically based, in tune with the organic chemist's intuition, and it is portrayable with computer graphics.

C. Acetylcholine Conformers

To complete this phase of the discussion on the dual-metric measuring of interatomic distances, we calculated the acetylcholine E-State values in their *gauche* and *trans* conformations. The coordinates were obtained from a standard drawing program. The interatomic distances were calculated among all nonhydrogen atoms. The calculation of the E-State values was based on the algorithm described previously.

The differences between the two conformations, shown in Fig. 10.3, produce differences in the E-State values calculated by this method. The *trans* conformation values reflect the maximum distances possible between the onium group atoms and the ester group atoms. When the conformation is folded into the *gauche* conformation, perturbations in certain atom E-State values occur. In the *gauche* conformer, the total loss of E-State value amounts to more than 0.1 unit. The majority of this loss is attributed to the ester group, in particular the ether oxygen atom. The atoms between these two groups are relatively unaffected by the change in conformation. The gain in E-State value by the ester group is a direct result of the electronegativity differences between the oxygen atoms and the onium group atoms. The transfer proceeds from the lower to higher electronegative atoms. This is in harmony with intuition.

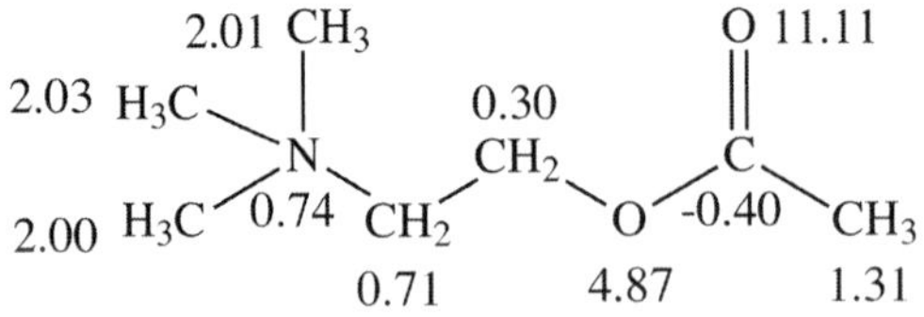

trans conformation

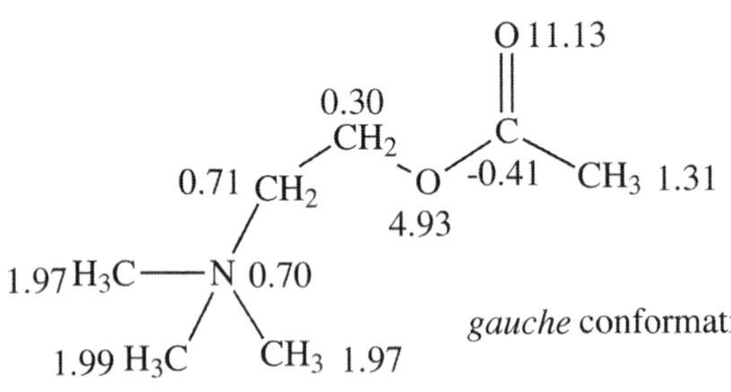

gauche conformation

FIGURE 10.3 Acetylcholine shown in both *trans* and *gauche* conformations along with the E-State values computed using Euclidean distances.

These results demonstrate many of the elements in a typical analysis using this paradigm. It can clearly be automated, beginning with a structure input or structure recall, followed by some energy minimization such as molecular or quantum mechanics. The interatomic distances may then be extracted and integrated with the graph distances to perform the E-State calculations. Further refinements are certainly in order, since the basic ingredients of a three-dimensional graph-based calculation are present. The objective of defining molecular structure and its influence on properties may be advanced a step further with this approach.

IV. PARAMETERIZING THE ONIUM GROUP

Quaternary nitrogen atoms in the sp^3 hybrid state frequently appear in molecules of biological interest. Consideration of this group should be included in some E-State studies in order to make the method as broadly applicable as possible. What is needed is a model which is in harmony with the other intrinsic state values and which permits an approximation of the intermolecular and intramolecular characteristics of the onium groups. A useful model should encode the observations that onium groups are quite electronegative, have varying topological status in a molecule, and are highly interactive across space. One proposed model (Kier and Hall, 1993) begins with the recognition that the ammonium group, $—NH_3^+$, is more electronegative than the amine group, $—NH_2$. This is true because the effective core charge is higher in the onium group. Accordingly, the representation of the core of the onium group must be defined to account for this difference. This definition must take into account the presence of a larger number of hydrogen atoms on the onium nitrogen relative to the amine group.

One approach is to regard the $—NH_3^+$ onium group as a composite group or pseudoatom made up of five valence electrons from the nitrogen atom and one electron each from two hydrogen atoms. The core of the pseudoatom is made up of the nitrogen atom core and the cores of the two hydrogens sharing sigma bond electrons. The third hydrogen atom remains outside of this pseudoatom model, sharing the nitrogen atom lone-pair electrons to form a sigma bond. This model of RNH_3^+ is shown in Fig. 10.4a. From the general equation for the intrinsic state, the I value is 7.0 as shown in Table 10.2. The pseudoatom model of a secondary onium pseudoatom is represented by Fig. 10.4b. In this model, there are six valence electrons resulting in $\delta^v = 5$ and two sigma bonds resulting in $\delta = 2$. The intrinsic state for the pseudoatom $[NH_2]$, from the equation for I, is determined to be $I = 3.0$ as shown in Table 10.2. The tertiary onium group is represented by the pseudoatom in Fig. 10.4c. Here, there are five valence electrons and three sigma bonds resulting in $\delta^v = 4$ and $\delta = 3$.

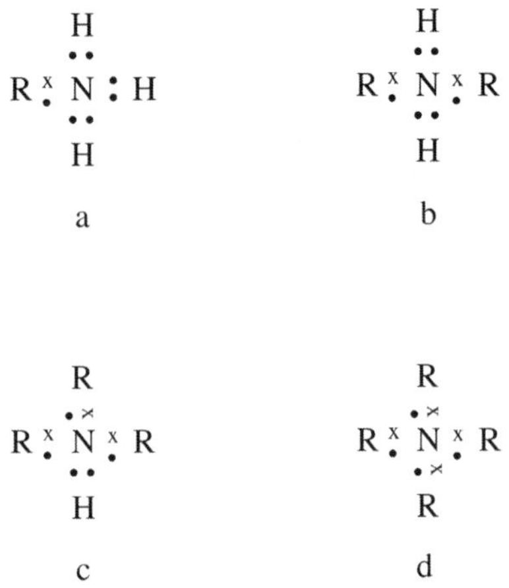

$$
\begin{array}{cc}
\text{H} & \text{H} \\
{\bullet\bullet} & {\bullet\bullet} \\
\text{R} \overset{x}{\underset{\bullet}{}} \text{N} \overset{\bullet}{\underset{\bullet}{}} \text{H} & \quad \text{R} \overset{x}{\underset{\bullet}{}} \text{N} \overset{x}{\underset{\bullet}{}} \text{R} \\
{\bullet\bullet} & {\bullet\bullet} \\
\text{H} & \text{H} \\
\text{a} & \text{b}
\end{array}
$$

$$
\begin{array}{cc}
\text{R} & \text{R} \\
{\bullet\,\times} & {\bullet\,\times} \\
\text{R} \overset{x}{\underset{\bullet}{}} \text{N} \overset{x}{\underset{}{}} \text{R} & \quad \text{R} \overset{x}{\underset{\bullet}{}} \text{N} \overset{x}{\underset{}{}} \text{R} \\
{\bullet\bullet} & {\bullet\,\times} \\
\text{H} & \text{R} \\
\text{c} & \text{d}
\end{array}
$$

FIGURE 10.4 Electron arrangements for the four onium groups according to the description in text. a, RNH_3^+; b, $R_2NH_2^+$; c, R_3NH^+; d, R_4N^+. The dots represent electrons associated with the pseudocore.

The intrinsic state calculated for this group, designated as [NH], is therefore $I = 1.67$.

In the final case, the quaternary nitrogen atom, the modeled pseudoatom, is represented by [N]. Here, there are five valence electrons resulting in a value $\delta^v = 5$ and four sigma bonds leading to $\delta = 4$. The intrinsic value calculated is $I = 1.5$ for this pseudoatom [N] (Table 10.2). This model succeeds, at least in a relative sense, in parameterizing these onium groups so that they reflect the electronegativity within a molecule and possibly their interactive capability with other molecules. It must be emphasized that these are provisional models that are devised to encode this structure feature in harmony with the other covalently bound atoms and hydride groups described previously. The parameters for these pseudoatoms are summarized in Table 10.2.

TABLE 10.2 Intrinsic States of Onium Pseudoatoms

Onium group	Onium pseudoatom	Z^v	δ^v	δ	I
RNH_3^+	$[NH_3]$	7	6	1	7.00
$R_2NH_2^+$	$[NH_2]$	6	5	2	3.50
R_3NH^+	$[NH]$	5	4	3	1.67
R_4N^+	$[N]$	5	4	4	1.50

V. E-STATE INDEX FOR BONDS AND BOND TYPES

It is important to be able to analyze molecular interactions in terms of the salient features of the interacting molecules. The atom-level E-State index has proven very useful in biological quantitative structure–activity relationship (QSAR) for this reason; it focuses on electron accessibility atom by atom. We now discuss a similar form of analysis based on the bonds of the molecule.

The local characteristics of an —OH group are very similar from molecule to molecule, especially in comparison with other groups such as —NH$_2$ or —CH$_3$. Nonetheless, the detailed character and behavior of an —OH group does depend on its immediate surroundings. Molecular structure representation may be improved by specifying the bond of which the —OH group is a part. For example, primary, secondary, and tertiary alcohols may require separate identification and characterization as a basis for good models. In a manner similar to that for the atom-type E-State, a value may be computed for each bond type in the molecule. We can expect the number of different bond types to be much greater than the number of atom types. The increased basis for molecular structure representation presents greater opportunity for development of satisfactory property prediction models, especially for large and heterogeneous data sets.

A. STEPS IN THE FORMALISM

In a manner parallel to the development of the atom-type E-State indices, we present the development for the bond type.

1. Classification of Graph Edges into Bond Types

An E-State value may be calculated for each bond in the molecule. Bonds may be classified into bond types in terms of the characteristics of the two atoms which define the bond. In the graph of 2-pentanol (Scheme III), there are several bonds and bond types:

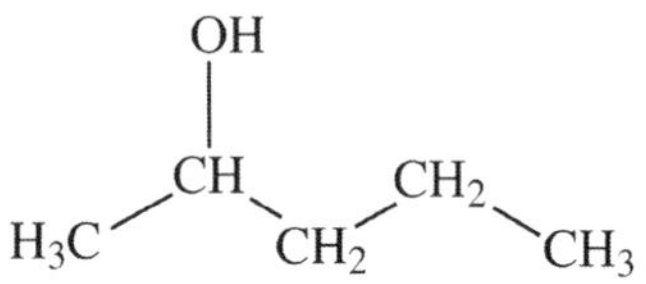

$$H_3C-CH \quad HC-OH \quad HC-CH_2 \quad H_2C-CH_2 \quad H_3C-CH_2$$

a   b   c   d   e

Each atom type can bond to many other atom types. In this example, the methine group, $>CH-$, is bonded to three different groups: a $-CH_3$, a $-CH_2-$, and an $-OH$. The only restriction is that each atom type must possess at least one of the same order of bond: single, double, aromatic, or triple. If there are n atom types, then, in general, there will be fewer than $n(n - 1)/2$ bond types because some combinations of atoms types do not occur, e.g., a bond between $-CH_3$ and $=O$. Classification may be based on an $n \times n$ matrix of the atom types. Any combination of atom types which does not correspond to an actual bond is eliminated. For example, the combination of $-CH_3$ and $=O$ atom types is an empty entry in the matrix of bond types. Currently, Molconn-Z computes E-State values for 52 atom types commonly found in organic molecules along with 450 bond types. (Atom and bond types for some inorganic species are also available but are not the topic of this discussion.)

2. Intrinsic State Value for Bonds

Our proposal for the intrinsic state value of a given bond, BI_s, is that BI_s for each bond, e_s, be obtained as the geometric mean of the atom I values for the two atoms in the bonds between atoms i and j:

$$BI_s = [(I_i I_j)]^{1/2} \qquad (10.8)$$

Based on this relation, intrinsic state values can be computed for each bond type in the molecule.

3. Perturbation Term for Bond Interaction

In the same manner as for atom-level E-State, we propose the perturbation on bond s by bond t be of the difference form

$$\Delta BI_{st} = BI_s - BI_t \qquad (10.9)$$

The perturbation is considered to increase with the difference between intrinsic states of the two bonds.

4. Bond E-State Index

As with the atom-level E-State, we consider the bond E-State value to be the perturbed intrinsic state; that is, the intrinsic state plus the sum of the perturbations by all other bonds in the molecule. The perturbation decreases with increasing distance between the two bonds. The distance is counted as the average atom–atom distance for the four atoms in the two bonds. The bond E-State value for a given bond, BS_s, is determined as follows:

$$BS_s = BI_s + \frac{\Sigma_t \, \Delta BI_{st}}{r_{st}^2} \tag{10.10}$$

The graph distance term r_{ij}, indicated in Scheme IV, can be obtained as the average of the atom–atom distance terms:

$$r_{st} = \frac{r_{is} + r_{it} + r_{js} + r_{jt}}{4} \tag{10.11}$$

(IV)

The software Molconn-Z computes the bond E-State index for each individual bond as well as for all the bond types in the molecule and makes them available for use in QSAR analyses.

A systematic symbolism has been developed to indicate each of the 450 organic bond types. Each symbol is of the following form: $eiAT_1AT_2$. For example, the bond order is indicated in ei as $e1$, $e2$, $e3$, and ea for single, double, triple, and aromatic bond orders. Each atom type consists of the element symbol and an indication of the number and type of bonds involving that atom type. Simple and familiar organic bond types can be symbolized readily. For example, for the CH_3—CH_2 bond, the symbol is $e1C1C2$; for the tertiary alcohol bond >CH—OH, it is $e1C3O1$; and for the ketone, it is $e2C3O1s$. The terminal "s" indicates that one of the atoms also has a single bond not used in this bond type. In this case, the single bond must be on the carbon atom since the keto oxygen has only one double bond. (These special symbols are only necessary in computer programming in which special characters, e.g., —, ═, and <, are not permitted.)

B. SAMPLE CALCULATION

The bond E-State and bond-type E-State values for the five bonds in 2-pentanol (Scheme IV) are given in Table 10.3. The upper part of the table also lists the atom types present in Scheme IV along with the atom E-State index values. The lower part of Table 10.3 lists each bond, the atom types of which it is comprised, the bond-level E-State value, and the bond-type E-State symbol. These are the bond-level E-State index values for the five bonds in 2-pentanol. The

TABLE 10.3   Bond E-State Values for 2-Pentanol

Atom ID	Atom group	E-State
1	—CH$_3$	1.80903
2	>CH—	−0.10185
3	—OH	8.55236
4	—CH$_2$—	0.93056
5	—CH$_2$—	1.08102
6	—CH$_3$	2.06222

Bond No.	Atom ID i	j	Atom types		Order	Bond E-State	Bond symbol
1	1	2	—CH$_3$	>CH—	1	1.242	e1C1C3
2	2	3	>CH—	—OH	1	5.090	e1C3O1
3	2	4	>CH—	—CH$_2$—	1	0.734	e1C2C3
4	4	5	—CH$_2$—	—CH$_2$—	1	1.199	e1C2C2
5	5	6	—CH$_2$—	—CH$_3$	1	1.735	e1C1C2

bond-type E-State values are the sum of the E-State values for bonds of the same type. In this example, all the bonds are unique. Thus, there are also five bond-type E-State values, each equal to the corresponding bond-level value.

To further illustrate the meaning of the bond E-State indices, several examples are given in Table 10.4. The first seven possess the same skeleton as 2-methyl pentane. The structure variation is confined to the terminal atom in the bond labeled "b" for the first five entries. In the first case, there are two bonds of the type CH$_3$—CH<. The bond-type E-State index for this bond type, e1C1C3, is the sum of these two values, 3.484. Entries 6 and 8 also possess two bonds of this type. Entry 6 (4-methyl-2-pentene) contains two bonds of the type CH$_3$—C═. The bond-type E-State index for this molecule is e1C1C3d = 3.696.

In the first five entries in Table 10.4, the electronegativity of the terminal atoms in bond b increases. It can readily be seen that the bond E-State value for b also increases. The interpretation of this trend is that the electron accessibility in this bond also increases. Such an observation is in agreement with general chemical intuition and experience.

Entries 2, 6, and 7 in Table 10.4 represent another small subseries of structures: variation of the location of the double bond. The double bond in entry 2 has the highest value (2.775), whereas the double bond in entry 6 has the lowest value (1.963). The electron accessibility for the double bond in entry 2 is judged the greatest, whereas that for entry 6 is the lowest, and entry 7 is in between.

TABLE 10.4 Bond E-State Indices for a Series of Related Molecules

No.	Structure	a	b	c	d	e
1		1.742	1.742	1.234	1.421	1.860
2	CH$_2$	1.813	2.775	1.336	1.287	1.789
3	NH$_2$	1.492	3.416	0.984	1.310	1.797
4	OH	1.242	5.090	0.734	1.199	1.735
5	F	0.992	6.763	0.484	1.088	1.672
6		1.848	1.848	1.963	1.711	1.730
7		1.609	1.609	1.546	2.157	2.079
8	OH	1.617	1.617	1.012	0.912	4.832
9	HO	1.174	1.174	5.396	0.594	1.663

Finally, entries 4, 8, and 9 in Table 10.4 comprise another small subset: variation of the C—OH bond from primary to secondary to tertiary. In this case, the tertiary OH bond has the highest bond E-State value, whereas the primary bond has the lowest value. There are two competing effects here. The primary C—OH bond is topologically more accessible but the tertiary bond has an even greater increase in electron density because of the three nearby groups of lower electronegativity.

These cases illustrate the manner in which bond E-State indices vary with structure variation. The meaning of the bond E-State values parallels that of the atom-type E-State values. Both atom type and bond type give a measure of electron accessibility within the molecule, an indication of the potential for interaction with another molecule. Bond-type indices also reflect both the presence or the absence of bond types in the structure as well as a relation to count of bond types.

C. APPLICATION OF BOND-TYPE E-STATE TO BOILING POINT

The bond-type E-State indices have not been used in any study published to date. For purposes of illustration, we reinvestigated the set of boiling point data for the 372 alkanes, alcohols, and chloroalkanes described in Chapter 9 and reported by Hall and Story (1997). There were six atom-type E-State indices used in the previously reported analysis based on a neural network.

For this data set there are nine bond types for carbon–carbon bonds (e1C1C2, e1C1C3, e1C1C4, e1C2C2, e1C2C3, e1C2C4, e1C3C3, e1C3C4, and e1C4C4) and four bond types involving the OH group (e1C1O1, e1C2O1, e1C3O1, and e1C4O1) and four representing the Cl group (e1C1Cl, e1C2Cl, e1C3Cl, and e1C4Cl).

These 17 indices were entered into the Neuralyst program and processed as described in Chapter 9 for the atom-type E-State indices. The learning rate was set at 0.8 and the momentum at 0.6. A 17:8:1 architecture was selected for the final run which ran for 12,000 epochs. The testing set was randomly selected at 21% (78 compounds) of the total data set.

The statistics for this best run are as follows: r^2: 0.996, 0.997, and 0.996 for the total set, training set, and testing set, respectively. Mean absolute error (MAE): 2.80, 2.74, and 3.01 for the total set, training set, and testing set, respectively.

When compared to the same statistics for the atom-type E-State analysis, it is clear that these results are significantly better. For the atom-type E-State analysis the MAE for the test set was 4.03. The bond-type E-States provide a 25% improvement. The plot of calculated boiling point versus observed boiling point indicates that the predicted values form an even tighter line as shown in Fig. 10.5. The value obtained for MAE is by far the lowest reported to date for such a data set.

Although this single example is limited, it does appear to us, based on our preliminary experience with a few other data sets, that the bond-type E-State indices will be very helpful for large data sets. Perhaps these indices will find their best use in the development of a model for predicting log P for which there are many thousands of reliable data available. There are, of course, other properties for which large data sets are being accumulated such as fathead minnow toxicity or toxicity to *Tetrahymena pyriformis*. Also, there are increasing numbers of data sets of chromatographic retention data. Investigations of these data sets are under way.

VI. FINALE

In this book we have developed a basis for a molecular structure description that is part of a larger paradigm for representation of molecular structure. Our purpose is the development of structure representation for understanding and modeling certain types of biological, chemical, and environmental prop-

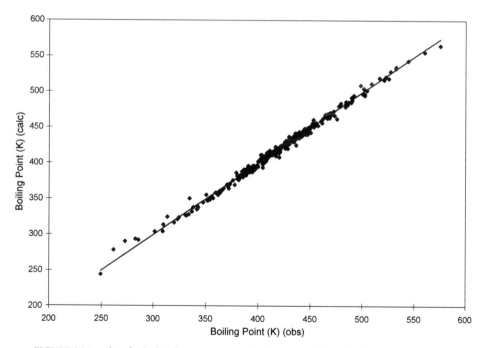

FIGURE 10.5 Plot of calculated versus observed boiling point (K) for 372 alkanes, alcohols, and polychloroalkanes from a model using a 17:8:1 artificial neural network based on bond-type E-State indice model.

erties. These properties have in common involvement in noncovalent inter-actions, molecular phenomena in which covalent bonds are neither broken nor formed. These events include the interaction of a drug with a receptor, the action of an inhibitor on an enzyme active site, interaction of dissolved solute molecules with water molecules, adsorption of a molecule on soil particles, va-porization of a substance, odor and taste of molecular substances, and others.

In these noncovalent interactions there are two central features of molecular structure that are essential for representation. For interaction there are elec-tronic factors, which may be described in terms of dispersion, dipolar, and hy-drogen bond forces. Each of these involves the arrangement of electrons within the molecule. The second factor is the topological accessibility of the electron distribution. When a ligand interacts with a receptor there are specific regions of the molecule that are complimentary to the receptor in a topological and electronic sense. Solubility is not as sensitive to complimentarity but is the re-sult of electron interaction as mediated by the molecular topology. The volatil-ity of a liquid is related to the electronic interactions among like molecules, interactions that are modulated by the degree of topological accessibility of the various electronic components of the molecule.

We have referred to the E-State index as a measure of electron accessibility for each atom. The more electronegative the atom, the greater the electron density around it, but the degree of topological accessibility at that atom modifies the potential for interaction with another molecule. Atoms on the mantle of the molecule have greater topological accessibility, while those buried within the molecule may have very little accessibility for interaction. In the E-State repre-sentation, electron richness is combined with the degree of topological accessi-bility. The heart of this formalism is the intrinsic state that combines electrone-gativity with local topology in a novel and useful basic manner. The possibility for intermolecular interaction depends upon electron richness but is moderated by the degree of topological accessibility. The perturbation of the fundamental intrinsic state by other atoms gives rise to the E-State value of an atom.

The concept of electron accessibility is part of a broader concept of accessi-bility that we have been developing. The molecular connectivity indices encode general topological accessibility. The $^1\chi$ index, for example, decreases with in-creased branching, reflecting the fact that some parts of the molecule are branch points, such as a quaternary carbon, and are consequently less available for contact with neighboring molecules. Seen in this light, the molecular connec-tivity chi indices directly encode topological accessibility, even for heteroatoms, whereas the chi indices only indirectly encode electron accessibility. On the other hand, the E-State indices directly encode both electron and topological accessibility. In addition, the extension to the hydrogen E-State provides a pa-rameter of representation of hydrogen accessibility.

The topological structure descriptors that we have developed provide the basis for a view of molecular structure based on the concept of accessibility. The array of molecular structure descriptors—molecular connectivity chi, kappa shape and E-State indices—gives the investigator a formidable arsenal of tools with which to attack a wide range of problems. It is entirely possible to use this structure mindset to model receptor binding, drug potency, toxicity, aqueous solubility, and others. In this manner, one may avoid the use of several different ways of thinking about structure, with a different method for each problem. The accessibility paradigm provides an integrated approach to modeling many different kinds of properties. The benefits of this approach are only just beginning to be seen on a broad scale.

VII. THE GRAND FINALE

The reader has come to the end of this book after being presented with concepts, descriptors, and applications. In one last statement, we briefly summarize the major take-home messages about the E-State paradigm that we would like the reader to remember.

A. DERIVATION

The E-State indices are nonempirical codes derived from the familiar chemical structure representation and the attributes of atoms and fragments in their bonding states within molecules. These attributes are the topology and the electronegativity of an atom in a molecule. The calculation is simple, rapid, unambiguous, and available for computer use.

B. INFORMATION

The indices encode the topological state of atoms and fragments plus their valence electron content. They encode the influence of each atom on the state of all other atoms in the molecule. These indices appear to be a significant improvement over the Hammett substituent concept. Unlike the Hammett constant, which encodes only the effect of a group through an aromatic ring to a reaction site, the E-State index encodes the influence of each atom, independent of its position, structure, or the context, over all other atoms in the molecule without limitation.

C. INTERPRETATION

The E-State indices encode the relative electron accessibility of an atom or fragment to engage in a putative intermolecular interaction forming an association of a noncovalent nature. These are the kinds of interactions found in drug–receptor encounters, enzyme–substrate inhibitions, and binding events common in molecular biological systems. They are also the kinds of interactions that dominate the manifestation of physical properties such as boiling points, solvent partitioning, and thermodynamic attributes.

D. ILLUMINATIONS

The E-State indices permit the illumination of salient features within molecules that are responsible for the phenomena at effector sites and at the available architecture of surrounding molecules in the neat state. The illumination of a pharmacophore is one example of the definition possibilities of the E-State indices.

E. GENERALITY

The atom-type E-State indices permit the successful structure–activity quantitation of noncongeneric series of molecules. These indices have also been shown to be the bases of field descriptors in comparative fields analyses (ComFA).

F. DATABASES

The E-State indices present a coherent pattern of numerical values that organize atoms, fragments, and molecules under any chemical description deemed of value by an investigator. This makes it possible to create databases, catalog molecules, and search and relate molecules according to any structural criteria of similarity. Concepts of similarity and diversity can be addressed with these indices, permitting a wide variety of applications.

VIII. A LOOK AHEAD

The next phase in the evolution of the E-State method lies in our hands as well as in those of investigators who have read this book or papers on the subject and who will benefit from its use. An accumulation of reports on the value of the E-State in various projects is anticipated as the value of the method is recognized in dealing with large data sets. We hope to hear from those who explore the method and who find success, discover shortcomings, or extend the method with useful innovations. We leave the reader with the hope for success and benefit from what we have presented in this book.

REFERENCES

Hall, L. H., and Story, C. T. (1997). Boiling point of a set of alkanes, alcohols and chloroalkanes: QSAR with atom type electrotopological state indices using artificial neural networks. *SAR QSAR Environ. Res.* **6**, 139–161.
Kier, L. B., and Hall, L. H. (1993). An atom centered index for drug QSAR models. *Adv. Drug Res.* **22**, 1–38.
Sanderson, R. T. (1976). *Chemical Bonds and Bond Energy*. Academic Press, New York.

INDEX

E-Calc

User's Guide

E-Calc

E-Calc is a program that calculates the Electrotopological State descriptors of molecular structure.

Academic Press   December 1998

Program Source:

Molconn-Z™

Hall Associates Consulting

2 Davis Street

Quincy, MA 02170

Authors:

SciVision and Lowell H. Hall

E-Calc Developer:

SciVision, Inc., 200 Wheeler Road, Burlington, MA 01803

Online Help:

Online help for topological indices supplied from SciQSAR-2D software by permission of SciVision, Inc., 200 Wheeler Road, Burlington, MA 01803. Text and graphics in online help developed by L. Mark Hall.

Contents

Introduction

E-Calc is a powerful utility that assists the user in calculating E-State values for molecules, including the electrotopological state (E-State) and hydrogen E-State (HE-State) values for individual atoms as well as the atom-type E-State indices. These calculations assist the reader in understanding the development, use, and interpretation of E-State values as a representation of molecular structure.

E-Calc enables the user to enter a molecular structure conveniently with the use of a drawing program. There is also a small library of stored structures that the user may call up. Scattered throughout the book are exercises that assist the reader in understanding the E-State formalism and the interpretation of the structure meaning of the E-State values. The user is encouraged to enter each of the structures described in the exercises and investigate the meaning of the values computed.

E-Calc computes all the E-State values and displays them in a convenient form on the screen. There is a convenient way to compare values for several molecules. There is no capability to import structures from other sources or to download the computed information for other use. The computational parts of this program have been taken from *Molconn-Z*™ [1] and from *SciQSAR*™*2D* [2]. For information on these and other related software packages, contact the vendors listed [1–4]. *E-Calc* is not designed for routine use in research nor for data sets of significant size or classroom use.

General Notes about Topological Descriptors

Topological descriptors of molecular structure may be computed from the connection table of a molecule. They may be computed for any structure that can be drawn or described in a connection table. No three-dimensional information is needed. Topological indices (structure descriptors) encode a wide range of structure information that has been used in the development of models that relate molecular structure to properties including biological, medicinal, physicochemical, and environmental [5–12].

There are three large classes of topological descriptors. The first set developed was the molecular connectivity chi indices [5–7]. The chi indices are whole molecule descriptors representing various aspects of the structure, especially related to degree of branching and cyclization.

The second set of descriptors is the kappa indices. These indices describe aspects of molecular shape such as cyclicity and central branching [8,9]. Molecular flexibility can also be encoded with the kappa indices.

The third set of topological descriptors is the E-State indices, which are described in the accompanying book [10–12]. These indices are computed for each atom in the molecule and represent a combination of both the electronic and topological states of that atom.

References

1. Molconn-Z™ software is available from Lowell H. Hall, Hall Associates Consulting, 2 Davis Street, Quincy, MA 02170 for DOS version only.

2. SciQSAR™2D is available from SciVision, Inc., 200 Wheeler Rd., Burlington, MA, 01803 for PC versions.

3. Molconn-Z is also available from EDUSOFT, LC., PO Box 1811, Ashland, VA, 23005 for UNIX versions of both a stand-alone type and a version interfaced to SYBYL™.

4. Molconn-Z is also available from Tripos, Inc., 1699 South Hanley Road, St. Louis, MO.

5. L. H. Hall and L. B. Kier, "Molecular Connectivity Chi Indices for Database Analysis and Structure-Property Modeling," in *Methods and Techniques for Structure-Property Modeling*, James Devillers, Alexandru T. Bababan, eds., Gordon and Breach (in press, 1998).

6. L. B. Kier and L. H. Hall, "Molecular Connectivity in Structure-Activity Analysis," Research Studies Press, John Wiley and Sons, Chichester, UK (1986).

7. L. B. Kier and L. H. Hall, "Molecular Connectivity in Chemistry and Drug Research," Academic Press (1976).

8. L. B. Kier, "Indexes of Molecular Shape from Chemical Graphs" in *Computational Chemical Graph Theory*, Chap. 6, pp. 152–174, D. H. Rouvray, ed., Nova Press, New York (1990).

9. L. B. Kier and L. H. Hall, "The Kappa Shape Indices for Modeling Molecular Shape and Flexibility," in *Methods and Techniques for Structure-Property Modeling*, James Devillers, Alexandru T. Bababan, eds., Gordon and Breach (in press, 1998).

10. L. B. Kier and L. H. Hall, "An Atom-Centered Index for Drug QSAR Models," in *Advances in Drug Research*, Vol. 22, pp. 1–38, B. Testa, ed., Academic Press (1992).

11. L. H. Hall and L. B. Kier, "The Electrotopological State: Structure Modeling for QSAR and Database Analysis," in *Methods and Techniques for Structure-Property Modeling*, James Devillers, Alexandru T. Bababan, eds., Gordon and Breach (in press, 1999).

12. L. B. Kier and L. H. Hall, "Molecular Structure Description: The Electrotopological State," Academic Press (1999).

Content of the Manual

Chapter One. Installation and Setup describes software and hardware requirements and how to install ***E-Calc.***

Chapter Two. Getting Started gives an overview of the ***E-Calc*** main window, menu items, and keyboard shortcuts.

Chapter Three. Building Molecules describes how to build a molecule in ***E-Calc***.

Chapter Four. Calculating Descriptors describes how to perform ***E-Calc*** calculations.

Appendix presents the list of descriptors.

Chapter 1. Installation and Setup

System Requirements

Computer

IBM, COMPAQ, Hewlett-Packard, NEC, or a true Pentium and Pentium II or compatible CPU. Digital Equipment NT Alpha computers can also run *E-Calc.*

RAM Memory

You must have at least 16 MB of random access memory (RAM).

Storage/Disk Drive

We recommend that you have a hard disk with at least 10 MB free. *E-Calc* files require approximately 5 MB.

Display

E-Calc supports video displays that are compatible with Microsoft Windows, including IBM Video Graphics Array (Super VGA). We strongly recommend a color monitor at 800×600 resolution.

Pointing Devices

You should have a Microsoft Mouse, or another mouse device compatible with Microsoft Windows 95/98.

Optional Hardware

You can use the MS Windows Control Panel to add and choose a printer if you want to print the results on a hard copy.

Software Requirements

You must have Microsoft Windows 95, Windows NT Version 4.0 or higher, or Microsoft Windows 98 installed on your computer before installing *E-Calc*.

Installing *E-Calc*

- Insert the CD-ROM in its drive.
- Select Start|Run.
- Select the CD-ROM drive.
- Select **Setup.exe** to install *E-Calc*.

E-Calc automatically installs and creates a new application group, *E-Calc*, with icons in the Start menu. One can drag the *E-Calc* icon from the *E-Calc* group dialog box. If it is missing, open up the Explorer window and select the *E-Calc* application icon. Drag it outside the Explorer window. An *E-Calc* shortcut icon will remain in the main startup window of Windows 95 or NT.

- Enter your Name, Company, and program Serial Number in the appropriate boxes during installation. Serial number is located on the program packaging.

For technical support call SciVision, the developers, at 1-800-861-6274 or e-mail SciVision at scivision@delphi.com.

Chapter 2. Getting Started

Starting *E-Calc*

Double click on the **E-Calc** icon to open up the main window.

Parts of the *E-Calc* Main Window

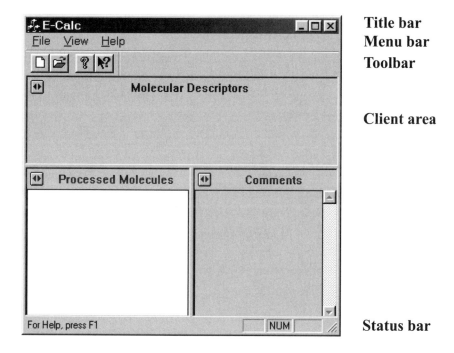

Title bar
Menu bar
Toolbar

Client area

Status bar

Client area...	Contains **Molecule Descriptors** pane, **Processed Molecules** pane, and **Comments** pane. Each pane can be enlarged by clicking the ◨ button.
Title bar...	Bears the title of the program, *E-Calc*, and the name of the current molecule.
Menu bar...	Contains *E-Calc* menus (**File, View, Help**).
Toolbar...	Displays available commands as icons in place of menu items.
Status bar...	The left area of the status bar describes actions of menu items as you use the arrow keys to navigate through menus.
	The right areas of the status bar indicate which of the following keys are latched down: Caps Lock (**CAP**), Num Lock (**NUM**), Scroll Lock (**SCRL**).

Menu Items

New...	Ctrl+N
Open...	Ctrl+O
Edit...	
Close	Ctrl+C
Close All	
1 0004.hin (1)	
2 0003.hin (1)	
3 0002.hin (1)	
4 0040.hin (1)	
Exit	

File menu commands

Commands of the File menu allow the user to create molecule files and to open and edit existing molecule files.

The underlined number commands enable the user to re-open (up to) four recently used files.

New	Opens the Sketcher window to build a new molecule.
Open	Invokes the Open dialog box to open the molecules stored in Library files.
Edit	Opens the Sketcher window to edit the current molecule.
Close	Closes the current molecule.
Close All	Closes all open molecules.
Exit	Terminates the current *E-Calc* session.

View menu commands

Commands of the View menu enable the user to set up the appearance of the main window and data presentation.

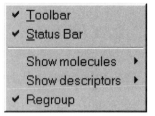

Toolbar	Shows/hides the toolbar.
Status Bar	Shows/hides the status bar.
Show molecules	Opens the popup menu with commands (**Tagged**, **All**) which specify that the **Molecular Descriptors** tables display data for either selected molecules or all molecules in the **Processed Molecules** list.
Show descriptors	Opens the popup menu with commands (**All**, **Non-zero**) which specify that the **Molecular Descriptors** tables display either all calculated descriptors or only the ones with nonzero values.
Regroup	Reformats the **Molecular Descriptors** tables, changing rows into columns.

Help menu commands

The Help menu contains the following commands common for all the windows of the program:

Contents   Displays the *E-Calc* Help Contents tab.

About E-Calc   Displays *E-Calc* version number and
                                       copyright notice.

Toolbar Icons

Toolbar icons replace the most important and frequently used *E-Calc* menu commands. Clicking on toolbar icons enables the user to perform the following commands:

Icon	**Command/Action**
	Opens the Sketcher window to create a new molecule.
	Invokes the Open dialog box to open molecule files.
	Displays *E-Calc* version number and copyright notice.
	Invokes the context-sensitive Help.

Shortcut Menus

In the *E-Calc* windows clicking the right mouse button invokes shortcut menus consisting of the most useful menu commands for the current window. Once opened, the shortcut menus are handled in the same way as common menus.

Keyboard Alternatives

E-Calc provides standard Windows alternatives to using the mouse. These alternatives are listed below:

♦ **To open or close a menu or a menu item using the keyboard alternative:**

- Each menu and item in the menu bar has an underlined letter. Press the <Alt> key and the <u>underlined</u> letter to open or select the menu, then the underlined letter to select an item.

- To close a menu or an item, press the <Alt> or <Esc> key.

Keyboard Shortcuts

E-Calc has few keyboard shortcuts. The following shortcuts perform the specific commands or replace the equivalent menu items:

- <F1>   Display the Help topic for the active window.

- <Esc>   Cancel a menu item.

- <Ctrl>+<N>   New on the File menu.

- <Ctrl>+<O>   Open on the File menu.

- <Ctrl>+<C>   Close on the File menu.

Chapter 3. Building Molecules

Sketcher Window

This chapter describes how *E-Calc* enables the user to build a new molecule. It describes the possibilities provided by the Sketcher window.

- Start *E-Calc* and select the New command on the File menu. This brings up the Sketcher window.

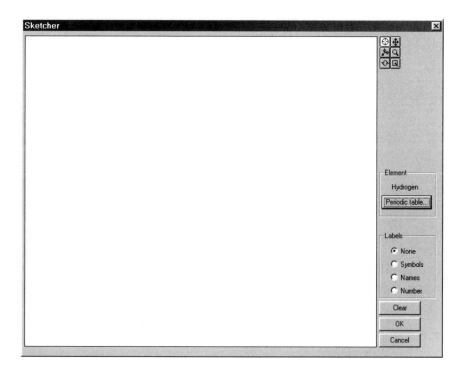

The Sketcher window is a tool for building molecules. It contains the build area and several tools to handle the building process. We will examine the possibilities of the window by building the ammonia molecule.

- Click the **Build** tool ⊕ in the upper right part of the Sketcher window.

- Select the **Names** option button in the **Labels** list.

- Click the **Periodic Table** button to invoke the Element Table dialog box:

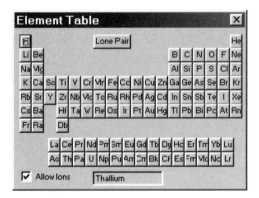

The Element Table dialog box enables the user to select an element to be drawn in the build area. The name of the selected atom is displayed in the text box below the table. Clicking on the **Lone Pair** button allows the user to draw a pair of electrons. Selecting the **Allow Ions** check box makes it possible to draw molecular fragments with no regard to valency limitations.

- Select **Nitrogen** and click inside the build area:

○

Nitrogen

- Select **Hydrogen** and click three times inside the build area:

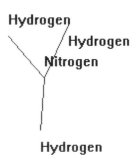

- To draw the bonds between the atoms, click on an atom, press the left mouse button, draw the mouse to the second atom, then release the mouse button.

Note: To create a double, triple, or aromatic bond, select the **Build** tool and consecutively click on the bond with the left mouse button to get the desired bond type. To delete a bond, just click on the bond with the right mouse button. To delete an atom, click on the atom with the right mouse button and select the Delete Atom command on the shortcut menu. To delete the whole molecule, just click the Clear button in the lower right corner of the Sketcher window.

Tools in the Sketcher Window

The Sketcher window provides the user with several tools to handle the molecules. We already saw the possibilities of the Build tool. Now we will examine the other tools.

Clicking on a tool opens a dialog box with controls for moving, rotating, zooming, etc., the molecules. At the same time the cursor acquires the shape of an image drawn on a tool. The respective actions can be performed by positioning the cursor inside the build area, pressing the left mouse button, and moving the cursor over the build area.

- Clicking on the **Move** tool ⊕ opens the Move dialog box with slide bars for moving the molecule over the build area in the specified directions. Clicking on **Reset** restores the initial position of the molecule.

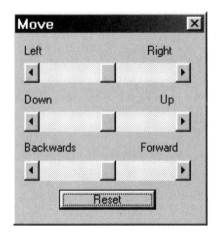

• Clicking on the **Rotate** tool 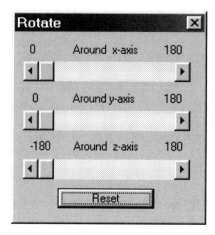 opens the Rotate dialog box with slide bars for rotating the molecule over three Cartesian axes.

• Clicking on the **Zoom** tool 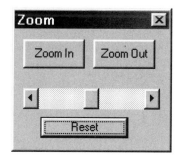 opens the Zoom dialog box with two buttons and slide bar for zooming the image in/out.

- Clicking on the **Plane** tool 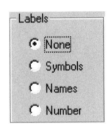 opens the Rotation in Plane dialog box with slide bars for rotating the molecule in the plane of the build area.

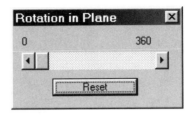

- Clicking on the **Restore** tool 🔲 restores the initial position of the molecule.

Changing Atom Labels

There are several types of labeling atoms in the Sketcher window that are set up by option buttons in the **Labels** frame. The user can display no label (**None**), the symbol of each chemical element (**Symbols**), the name of each element (**Names**), or ID number of each atom (**Number**).

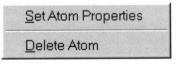

- Click each option button to view changes in the display of the labels.

Shortcut Menu

On selecting the Build tool and clicking on a particular atom in the build area with the right mouse button the shortcut menu appears. The menu consists of two commands.

Set Atom Properties

Delete Atom

The Delete Atom command deletes the selected atom from the build area as described above.

The Set Atom Properties command opens the Atom Properties dialog box:

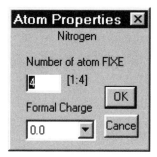

The dialog box contains two controls that enable the user to change the number of the atom (as displayed by selecting the **Number** option in the **Labels** section) and to set up the formal charge of the atom.

Closing the Sketcher Window

On closing the Sketcher window *E-Calc* performs the calculation of the descriptors and displays the result in the main window.

- Click **Close** to open the main window with the calculation results for the ammonium molecule. We will discuss the presentation of the calculation results in the following chapter.

Chapter 4. Calculating Descriptors

Introduction

This chapter describes the presentation of the calculated results in the *E-Calc* main window. It will cover both calculations of single molecules and batch mode calculations for molecules stored in files.

Calculations for Molecules Built in *E-Calc*

As it was described at the end of the previous chapter, after building a molecule in the Sketcher window and closing the window, *E-Calc* automatically computes all the E-State values and enables display of calculated results in the main window.

We will study in detail the features of the main window in the next section.

Calculations for Molecules Stored in Files

E-Calc enables the user to perform the calculations for molecules stored in files. Both calculations for a single molecule and batch-mode calculations for multiple molecules are available.

Opening Multiple Files

- Start *E-Calc* and select the Open command on the File menu or click the respective toolbar button.

This brings up the Open dialog box.

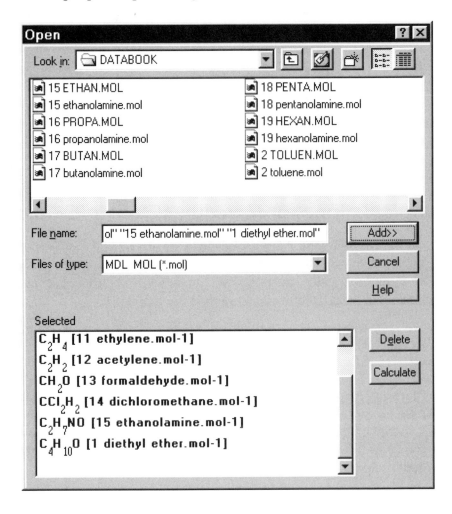

This is an extended version of the standard Windows Open File dialog box. Files of the specified types can be selected in the folder list by clicking on a file name. To select consecutive files press the <Shift> key and click on the first and the last file name; to select several files press the <Ctrl> key and click on the file names desired. Clicking on **Add** moves files selected in the folder to the **Selected** list. Clicking on the file in the Selected list highlights the file, and the file can be removed from the list by clicking on the **Delete** button.

- Select the files as shown in the above picture and click on the **Calculate** button to perform the calculations and display the calculations result in the *E-Calc* main window.

Presentation of the Calculation Results

On performing the batch-mode calculation as described above, the main window will look like:

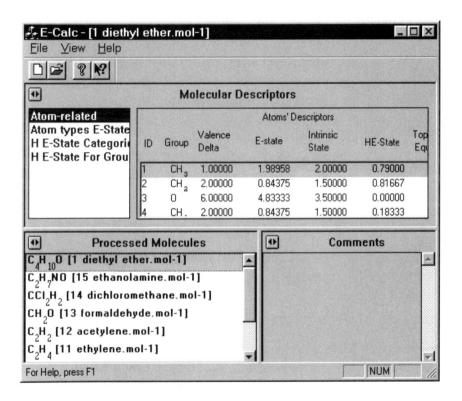

The main window displays the data in three panes.

Molecular Descriptors pane

The Molecular Descriptors pane displays the calculated descriptor values for the processed molecules. The descriptors are split into four groups and are thus presented in four tables: **Atom-related, Atom types E-State Indexes, H E-State Indexes Categories**, and **H E-State For Groups**. The user can switch between the tables by clicking on the respective group in the left-side list of the pane.

The shortcut menu of the Molecular Descriptors pane contains the following commands:

Show molecules Opens the popup menu with commands (**Tagged, All**) which specify that the **Molecular Descriptors** tables display data for either selected molecules or all molecules in the **Processed Molecules** list.

Show descriptors Opens the popup menu with commands (**All, Non-zero**) which specify that the **Molecular Descriptors** tables display either all calculated descriptors or only the ones with nonzero values.

Regroup Reformats the **Molecular Descriptors** tables changing rows into columns.

Processed Molecules pane

The Processed Molecules pane contains the list of processed molecules. A molecule can be selected (highlighted) by clicking on a molecule in the list. To select consecutive molecules, press the <Shift> key and click on the first and the last molecule. To select several molecules, press the <Ctrl> key and click on the desired molecules. The shortcut menu of the Processed Molecules pane contains the following commands: **Edit** (opens the selected molecule in the Sketcher window), **Close** (removes the molecule from the list together with the data in the Molecular Descriptors pane), and **Close All** (clears all panes in the main window).

Comments pane

The Comments pane displays the respective text for molecule file formats containing comments. A fragment of the text can be selected with the mouse. The Comments pane has a shortcut menu with the following commands: **Undo**, **Cut**, **Copy**, **Paste**, and **Select All**, which enable the user to export text fragments or the whole text to/from the Clipboard.

Molecular Descriptors Calculated in *E-Calc*

E-Calc computes the E-State value and the hydrogen E-State value for each individual atom in the molecule. These individual atom descriptors are displayed for each atom in a single screen along with information on the valence delta value, the Kier–Hall electronegativity, and symmetry information.

In addition to the atom E-State values, **E-Calc** provides the atom-type and hydrogen atom-type E-State values. Below is a table of the symbols used for both atom-type E-State indices.

An atom-type E-State index is the *sum of E-State values* for all atoms of the given type in the molecule.

Table 1

Information for Atom-Type E-State Indices

No	Atom group[a]	Z[b]	δ^v[c]	δ[d]	Valence state indicator $\delta^v + \delta$[e]	$\delta^v - \delta$[f]	AR[g]	Atom-type E-State Symbol[h]
1	-Li	3	1	1	2	0	0	SsLi
2	-Be-	4	2	2	4	0	0	SssBe
3	>Be<⁻	4	2	4	6	-2	0	SssssBem
4	>BH	5	3	2	5	1	0	SssBH
5	>B-	5	3	3	6	0	0	SsssB
6	>B<⁻	5	3	4	7	-1	0	SssssBm
7	-CH₃	6	1	1	2	0	0	SsCH3
8	=CH₂	6	2	1	3	1	0	SdCH2
9	-CH₂-	6	2	2	4	0**	0	SssCH2
10	≡CH	6	3	1	4	2**	0	StCH
11	=CH-	6	3	2	5	1	0	SdsCH
12	--CH--	6	3	2	5	1	1	SaaCH
13	>CH-	6	3	3	6	0**	0	SsssCH
14	=C=	6	4	2	6	2*	0	SddC
15	≡C-	6	4	2	6	2*	0	StsC
16	=C<	6	4	3	7	1	0	SdssC
17	--Ċ--	6	4	3	7	1	1	SaasC
18	--Ċ--	6	4	3	7	1	1	SaaaC
19	>C<	6	4	4	8	0	0	SssssC

20	-NH$_3^+$	7	2	1	3	1	0	SsNH3p
21	-NH$_2$	7	3	1	4	2	0	SsNH2
22	-NH$_2^+$-	7	3	2	5	1	0	SsNH2p
23	=NH	7	4	1	5	3	0	SdNH
24	-NH-	7	4	2	6	2	0	SssNH
25	--NH--	7	4	2	6	2	1	SaaNH
26	≡N	7	5	1	6	4**	0	StN
27	>NH$^+$-	7	4	3	7	1	0	SssNHp
28	=N-	7	5	2	7	3	0	SdsN
29	- -NH- -	7	5	2	7	3	1	SaaN
30	>N-	7	5	3	8	2*	0	SsssN
31	$-\overset{\shortparallel}{N}=$	7	5	3	8	2*	0	SddsN (nitro)
32	$--\overset{\shortparallel}{N}--$	7	5	3	8	2*	0	SdaaN (N oxide)
33	>N$^+$<	7	5	4	9	-1	0	SssssNp (onium)
34	-OH	8	5	1	6	4	0	SsOH
35	=O	8	6	1	7	5	0	SdO
36	-O-	8	6	2	8	4	0	SssO
37	- -O- -	8	6	2	8	4	1	SaaO
38	-F	9	7	1	8	6	0	SsF

39	-SiH3	14	4	1	5	3	0	SsSiH3
40	-SiH2-	14	4	2	6	2	0	SssSiH2
41	>SiH-	14	4	3	7	1	0	SsssSiH
42	>Si<	14	4	4	8	0	0	SssssSi
43	-PH$_2$	15	3	1	4	2	0	SsPH2
44	-PH-	15	4	2	6	2	0	SssPH
45	>P-	15	5	3	8	2	0	SsssP
46	$-\!\!\!\!\stackrel{}{P}\!=$	15	5	4	9	1	0	SdsssP
47	$-\!\!\!\!\stackrel{}{P}\!\!<$	15	5	5	10	0	0	SssssssP
48	-SH	16	5	1	6	4	0	SsSH
49	=S	16	6	1	7	5	0	SdS
50	-S-	16	6	2	8	4	0	SssS
51	- -S- -	16	6	2	8	4	1	SaaS
52	>S=	16	6	3	9	3	0	SdssS (sulfone)
53	$-\overset{\|\|}{\underset{\|\|}{S}}-$	16	6	4	10	2	0	SddssS (sulfate)
54	$\overset{\|}{\underset{\|}{S}}$	16	6	6	12	0	0	SssssssS
55	-Cl	17	7	1	8	6	0	SsCl

56	-GeH3	32	4	1	5	3	0	SsGeH3
57	-GeH2-	32	4	2	6	2	0	SssGeH2
58	>GeH-	32	4	3	7	1	0	SsssGeH
59	>Ge<	32	4	4	8	0	0	SssssGe
60	-AsH2	33	5	1	6	4	0	SsAsH2
61	-AsH-	33	5	2	7	3	0	SssAsH
62	>As-	33	5	3	8	2	0	SsssAs
63	->As=	33	5	5	10	0*	0	SdsssAs
64	->As<	33	5	5	10	0*	0	SssssssAs
65	-SeH	34	5	1	6	4	0	SsSeH
66	=Se	34	6	1	7	5	0	SdSe
67	-Se-	34	6	2	8	4	0	SssSe
68	..Se..	34	6	2	8	4	1	SaaSe
69	–S̈e–	34	6	3	9	3	0	SdssSe
70	–S̈e–̈	34	6	4	10	2	0	SddssSe
71	-SnH3	50	1	1	2	0	0	SsSnH3
72	-SnH2-	50	2	2	4	0	0	SssSnH2
73	>SnH-	50	3	3	6	0	0	SsssSnH
74	>Sn<	50	4	4	8	0	0	SssssSn
75	-Br	35	7	1	8	6	0	SsBr

76	-I	53	7	1	8	6	0	SsI
77	-PbH3	82	1	1	2	0	0	SsPbH3
78	-PbH2-	82	2	2	4	0	0	SssPbH2
79	->PbH-	82	3	3	6	0	0	SsssPbH
80	>Pb<	82	4	4	8	0	0	SssssPb

* Denotes an atom which can be distinguished from another only by analysis of neighboring atoms.

** Cases in which the delta difference, $\delta^v - \delta$, is required as an additional classification criterion.

[a] Indication of atom groups, according to valence state, together with number of bonded hydrogens.
Symbols: - for single, = for double, ≡ for triple, -- for aromatic.

[b] Atomic number of element.

[c] Valence connectivity delta value.

[d] Simple connectivity delta value.

[e] Sum of valence and simple delta value.

[f] Difference of valence and simple delta value.

[g] Indicator that atom is part of aromatic system: 1 indicates aromaticity.

[h] Symbol used for atom types in early publications.

Table 2

Atom Types and Group Types Used with the Hydrogen Atom-Type E-State Indices

A hydrogen atom-type E-State index is a *sum of the hydrogen E-State values* for all atoms of the given type in the molecule.

A. Hydrogen Atom-Type

	Types	Symbol	Comments
1	-OH	SHsOH	
2	=NH	SHdNH	
3	-SH	SHsSH	
4	$-NH_2$	SHsNH2	
5	-NH-	SHssNH	
6	$\equiv$CH	SHtCH	
7	$-CH_nX$	SHCHnX	a CH or CH_2 group with a -F or -Cl also bonded to the carbon
8		SHother	the sum of E-State values for all groups not listed above

B. Maximum and Minimum Values

	Symbol	Comments
9	Hmax	maximum hydrogen E-State value in molecule
10	Gmax	maximum E-State value in molecule
11	Hmin	minimum hydrogen E-State value in molecule
12	Gmin	minimum E-State value in molecule

A group-type hydrogen E-State index is the *sum of hydrogen E-State values* for all groups of the same type in the molecule.

C. Group Types

	Group	Symbol	Comments
13	-CH$_3$	SHtmet	
14		SHCsat	sum over all saturated carbon hydride groups in molecule: -CH$_3$, -CH$_2$-, >CH-.
15	=CH$_2$	SHtvin	
16	=CH-	SHvin	
17	≡CH	SHtrip	
18	..CH..	SHarom	sum over all aromatic CH groups

Subject Index

Symbols and Terms with Examples

Atom Type
Atom class based on valence state and number of bonded hydrogens (p. 67)

Atom Type E-State Value
Sum of the E-State values for all atoms of the same type in the molecule; symbol: e.g., $S^T(-CH_3)$, $S^T(=O)$, etc. (p. 67)

{Illustration}

Atom Type Sum E-State Values	**Atom Type Symbols**	**Atom Type Sum H E-State Values**
$S^T(-OH)$ = 17.88	OH -OH	$HS^T(-OH)$ = 5.08
$S^T(--CH--)$ = 4.67	--C-- CH$_3$ -CH$_3$	$HS^T(--CH--)$ = 3.67
$S^T(-CH_3)$ = 1.67	--CH-- --C--	$HS^T(-CH_3)$ = 0.67
$S^T(--C--)$ = 0.78	--CH--	
	--CH-- OH -OH	

Delta Values

simple delta, δ the number of bonded neighbors in the molecular skeleton, (exclusive of hydrogen atoms); the sigma electron descriptor (p.18)

$\delta = \sigma - h$

σ = count of electrons in sigma orbitals
h = the count of bonded hydrogen atoms.

valence delta, δ^v the number of valence electrons in an atom, exclusive of hydrogen atoms; the valence electron descriptor (p. 18)

$\delta^v = Z^v - h = \sigma + \pi + n - h$

π = number of electrons in pi orbitals
n = number of electrons in lone pair orbitals

{Illustration}

Group	Z^v	σ	π	n	h	δ	δ^v	N	RKHE	I
-CH$_3$	4	4	0	0	3	1	1	2	0.00	2.00
=C<	4	3	1	0	0	3	4	2	0.25	1.67
-NH-	5	3	0	2	1	2	4	2	0.50	2.50
-OH	6	2	0	4	1	1	5	2	1.00	6.00
=O	6	1	1	4	0	1	6	2	1.25	7.00
-F	7	1	0	6	0	1	7	2	1.50	8.00
-S-	6	2	0	4	0	2	6	3	0.44	1.83
-Cl	7	1	0	6	0	1	7	3	0.67	4.11
-Br	7	1	0	6	0	1	7	4	0.38	2.75

Electronegativity, Kier-Hall (Relative)
Kier-Hall valence state electronegativity, relative to KHE[C(sp^3)] taken as zero (p. 19)

$KHE = (\pi + n)/N^2$

N is principal quantum number of valence electrons

Electrotopological State Value (E-State)
The E-State value calculated for each atom in the molecule, S(i)
For the E-State: (p. 23)
$$S(i) = I_i + \sum_j \Delta I_{ij}$$
For the Hydrogen E-State: (p. 61)
$$HS(i) = \sum_j \Delta HI_{ij} + ((-0.2) - KHE_i)$$

{Illustration}

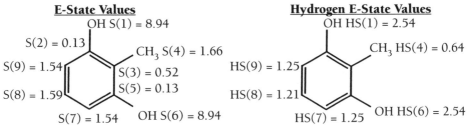

E-State Values	**Hydrogen E-State Values**
OH S(1) = 8.94	OH HS(1) = 2.54
S(2) = 0.13	
CH$_3$ S(4) = 1.66	CH$_3$ HS(4) = 0.64
S(9) = 1.54	HS(9) = 1.25
S(3) = 0.52	
S(8) = 1.59	HS(8) = 1.21
S(5) = 0.13	
S(7) = 1.54 OH S(6) = 8.94	HS(7) = 1.25 OH HS(6) = 2.54

Hydrogen-Suppressed Graph
Graphical representation of molecule structure; graph is set of objects (vertices) and the relations among them (vertices) (p. 15)

{Illustration}

$$H_2NC_6H_4COOCH_2CH_3 \Longrightarrow$$

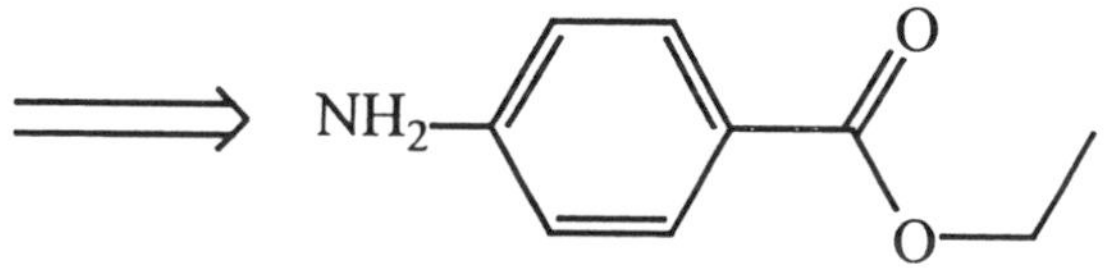

NH$_2$—

Intrinisic State Value
Electronic and topological character of valence state atom
For the E-State: (p. 21)
$$I = [(2/N)^2 \delta^v = 1]/\delta$$
  N = principal quantum number for valence electrons
For the Hydrogen E-state: (p. 61)
$$I = 0$$

{Illustration}

E-State Intrinsic State Values

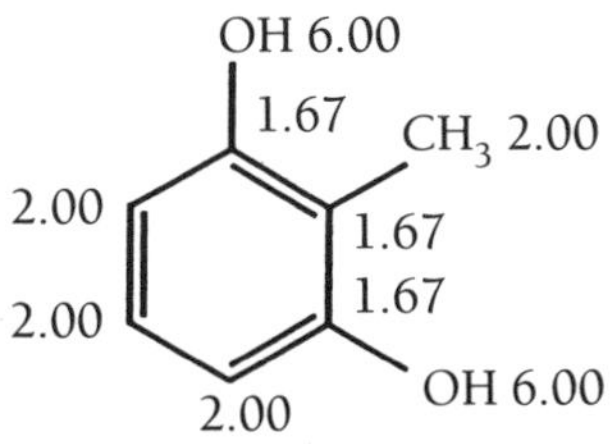

```
        OH 6.00
           1.67   CH_3 2.00
   2.00
                  1.67
   2.00           1.67
        2.00    OH 6.00
```

Perturbation Term
Electronic and topological influence of all other atoms on a given atom
For the E-State: (p. 22)
$$\Delta I_{ij} = (I_i - I_j)/r_{ij}^2.$$
  r_{ij} = number of atoms in the shortest path connecting atoms i and j
For the Hydrogen E-State: (p. 61)
$$\Delta HI_{ij} = (KHE_i - KHE_j)/r_{ij}^2$$